CW01501977

A. E. DOLBEAR

THE ART OF PROJECTING

A MANUAL OF EXPERIMENTATION
IN
PHYSICS
CHEMISTRY, AND NATURAL HISTORY
WITH THE
PORTE LUMIERE AND MAGIC LANTERN

Elibron Classics
www.elibron.com

Elibron Classics series.

© 2005 Adamant Media Corporation.

ISBN 1-4021-0506-1 (paperback)
ISBN 1-4021-0505-3 (hardcover)

This Elibron Classics Replica Edition is an unabridged facsimile
of the edition published in 1877 by Lee and Shepard,
Boston.

THE ART OF PROJECTING.

A Manual of Experimentation

IN

PHYSICS,

CHEMISTRY, AND NATURAL HISTORY

WITH THE

PORTE LUMIERE AND MAGIC LANTERN.

BY PROF. A. E. DOLBEAR,
TUFT'S COLLEGE.

ILLUSTRATED.

BOSTON:
LEE & SHEPARD, PUBLISHERS.
NEW YORK: CHARLES T. DILLINGHAM.
1877.

PREFACE.

THE object of this treatise is to point out to teachers of physical science, and to others who may be interested in experimentation, the usefulness of the Magic Lantern, and especially of the Porte Lumière, and a few other pieces of apparatus which can mostly be extemporized. With these a surprisingly large number of experiments in every department of physics may be performed, and every department of science and art may be illustrated; and the illustrations may be upon a scale of magnitude which will surprise one who has never witnessed them. The manipulation of the apparatus is not at all difficult, and no one need fear he will not succeed in doing anything described in the book, provided that at first he masters the simple conditions of projection with a single lens and with a condenser.

The simplest fixtures have been described, and a cut has been inserted wherever it could make more intelligible either the forms of the apparatus or the necessary conditions. No attempt has been made to explain phenomena, — other books do that; but it is hoped that a sufficient number and variety of experiments are plainly described to make any one thoroughly familiar with the art of projecting.

INDEX.

Absorption spectra 114
Acoustic curves 61
Air thermometer 144
Animalcule cage 33

Biaxial crystals 132
Bubbles 107
Calorescence 149
Camphor on water 47
Camera obscura 80
Candle power 13
  " flame, To project . . 92, 100
Capillarity 49
Caustics by reflection 92
  "   " refraction 104
Chameleon top 143
Chemical tank 34
  " reactions 157
Chladni's experiment 62
Chromatic aberration 104
Chromatrope 142
Cloud formation 145
Cohesion 45
Cohesion figures 47
College lantern 41
Colors of thin films 107
Concave mirror, To project
  with 63, 91
Convection in water 156
  "   " air 98
Condenser: its use 26
Convex mirrors 93
Crove's apparatus 77
Crystalline substances for polar-
  ized light 133

Darkened room 5
Diagrams on mica 129
Diamagnetism 151
Diffraction 137
Disks for study of colors . 110, 143
Dispersion 105
Distortion 93
Divisibility of matter 44
Double refraction 126
Double salts, Prepared . . . 134
Drummond light 11

Eidotrope 42
Electric light 9
  "   " To project . . . 153
Engravings, To transfer . . . 82
Etching upon glass 31

Fluorescence 119
Focal length of lenses 21
Focusing 25
Fountain, Illuminated 96
Fraunhofer's lines 111

Galvanometer 147
Gases for lime light 11
Ghost 84
Glue, Marine 35
Gramme machine 9
Gravitation 50

Heat 144, 155
Heliostat 1

Ice flowers 52
Illumination, Intensity of . . 81
Images formed by lenses . . 100
Interference 71
  "   spectra 118
Interlacing lines 70

Kaleidoscope 88
Kaleidophone 57

Lanterns 14
Lenses 19
  " Magnifying power . . . 33
  " Mountings for 23
Light 80
  " Intensity of 13
  " Magnesium 10
  " Lime 11
  " Composition of . 109, 117, 136
  " Polarized 127
Lissajou's experiments 69

Mach's experiment 64
Magnetism 150
Magnetic phantom 150
Manometric flames 62
Marine glue 35
Megascope 38
Melde's experiment 58
Microscope solar 100
  " attachment 49
Minute substances 133
Mirage 95
Monochromatic light . . . 108, 122

Newton's disk 143
  " rings 109

Objective 25
Objects for projection 27
Organ pipe 65
Opeidoscope 59
Outline drawings 29
Overtones 71

Persistence of vision 139
Pepper's ghost 84
Plateau's (experiment) 56
Polarization of light 127
Porosity 45
Projection with single lens . . 24
"   "   condenser . . . 27
"   of large apparatus . 35
"   Apparatus for verti-
cal 40
Porte Lumiere, To make . . . 2
"   "   its use 24
Pyrometer 145

Rainbow 100
Reactions, Chemical 157
Reflections 82
"   Multiple 83
Refraction 97
Resultants 72

Salicine crystals 134
Screens 6
Sciopticons 18
Silver crystals 53
Singing flames 64

Sinuous lines 69
Soap bubbles, Persistent . . . 108
"   "   Tension of . . . 107
Solar microscope 100
"   spectrum 111
Spectacle glasses, To test . . . 132
Spheroidal form 62
Spectra, Methods of project-
ing 121, 153
Spectrum analysis 119
"   of sodium 121
"   "   "   reversed . 122
Starch 134
Stroboscope 139
Sympathetic vibrations 75

Thermometer 144
Total reflection 94
Tuning forks 57

Vibrations of strings 59
"   "   forks 57
Vision, Persistence of 139

Water, Decomposed 153
"   Maximum density . . . 146
"   Refraction of 97
"   Total reflection in . . 94
Waves in water 61
Whirling-table attachment . . 77

Zoetrope 140

THE ART OF PROJECTING.

A MAGNIFIED image of a picture, or of any phenomenon, when thrown upon a screen by means of sunlight,
and lenses, or with a magic lantern, is called a projection.

When sunlight is to be used for this purpose, it is
necessary to have some fixture to give the proper direction to the beam. The *heliostat* and the *porte lumiere*
are the devices in common use. The latter was the
earliest form, and was invented by Gravesand, a Dutch
professor of natural philosophy, in the early part of the
last century. It was afterwards reinvented by Captain
Drummond, an Englishman, who called it the *heliostat.*
The latter term is now only applied to an automatic
arrangement, by which a mirror is moved by clock-
work in such a way that a beam of sunlight reflected
from it may be kept in one direction all day, if it be
needed so long. Silberman and Foucault have each
devised very satisfactory instruments, but they are too
costly to be owned by any but the wealthy; the catalogue
price of the cheapest of these being five hundred
francs. C. Gerhardt, of Bonn, however, makes a small
one, carrying a good mirror three inches in diameter,
for twenty dollars.

THE PORTE LUMIERE — HOW MADE.

The *porte lumiere* is made of various patterns, and its movements are directed by turning milled-head screws. Ritchie makes an excellent one with three and a half inch aperture, for about twenty-five dollars, and it is recommended that such an one be purchased at the outset, if it can be afforded, but as many who would be glad to work with one cannot purchase it, directions will be given for making one, that will enable any person who is familiar with the use of carpenters' tools, to make one at a trifling cost that will answer every purpose.

The room in which the *porte lumiere* is to be used must, of course, be one into which the sun can shine. A room having windows only upon the North side, evidently cannot be used at all for such a purpose ; one having windows only upon the East or upon the West side could be used only in forenoon or afternoon ; while one with windows looking to the South can be used nearly all day. Choose then that window where the sun is available the longest, and opposite to which can be stretched the screen to receive the projections upon. Next, take a well-seasoned piece of pine board a foot or more in width, and an inch thick when dressed ; cut it to the length of the width of the window sash, so that it may fit into the window frame, and the sash be brought down upon it ; this will keep it tightly in place. With the compasses, scratch two concentric circles in the middle of the board, one with a radius of four inches, the other with a radius of four inches and a half. Saw out the inner circle completely, and cut the other but one half through the board, and then cut away, making a square rabbet, as shown at *b b*. Next, take a round piece of inch board of the same diameter as the outer circle (namely, nine inches), cut a rabbet upon one side of it so that it will nicely fit into the hole of the larger board, as indicated at *c c.*

Make the worked edges, and touching surfaces, quite smooth; but the outer edge should be made a trifle smaller than the hole, in order to allow the disk to turn freely round in it; then the hole may be cut in the disk to receive the lens, four or five inches in diameter, whichever it may chance to be.

Procure a nice piece of thin looking-glass, twelve or fifteen inches long and five inches wide. Fasten it to a back of wood made a little larger than itself, with broad-headed tacks, or bits of wire driven in and the top bent at right angles. This back will need to be an inch thick at the bottom, but may taper like a shingle to the top, where it need not be half an inch thick; *m* is the mirror and *h* is the back in the figure adjoining.

A common desk hinge *h* may be used to attach this mirror-mounting to the part *c* in the figure below. It must be so fastened that the mirror may swing through ninety degrees from a horizontal plane. The accompanying figure will be sufficiently definite to enable any one to make the whole instrument. When the mirror is securely fastened to the part *c*, the whole can be inserted in the board *b b* and *buttoned* in, as is shown at *b* and *b;* these buttons must

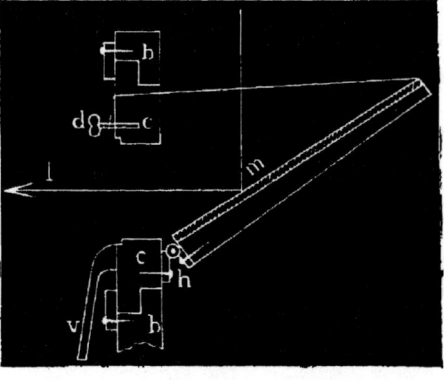

not bind upon the part *c*, as this must have an easy rotation in its place, though they need to be tight in the board *b;* three of them will be enough. Again, a string must be attached to the end of *m*, passed through a small hole in *c*, and tied to a tight-fitting thumb peg at *d.* As the peg is turned the mirror will be raised or lowered. A short lever *v* must be made fast to some part of *c* with which to turn the whole fixture around as the sun moves. The ray of light *l* can then be always kept where it is wanted.

If the window-sill be no more than two or three feet from the floor, it will be better to have this fixture either put into a window shutter, or to remove a pane of glass at the proper place and fasten the board *b b* into it. In this case it will be necessary to have a cap to place over the hole when it is not in use.

The lenses will need to be purchased; and for a beginning I recommend a cosmorama lens five or six inches in diameter and with a focal length of eighteen or twenty inches; a plano-convex lens of two and a half or three inches diameter and eight or nine inch focus; also a pocket botanical glass with focus of one or two inches. These three lenses should cost no more than six dollars, if the two former are unmounted. If one has got a magic lantern, or a sciopticon, the lenses in that will answer admirably. Take one of the glasses of the compound condenser and fasten it into the orifice of the *porte lumiere* with its convex side out; then, taking out the front lenses, hold them with one hand in the path of the divergent beam of light from the *porte lumiere,* and distant but four or five inches from it, and with the other hand hold some object between it and the larger lens; by moving the lens or the object a little, a sharp outline of it will be observed upon the opposite wall, and then will be seen what further conveniences will be wanted, such as curtains, screen, table, mounting for lenses, etc.

THE DARKENED ROOM.

Exhibitions with the stereopticon are almost always given at night, and there is no trouble from exterior light; but the illustrations and demonstrations which are part of the work of schools and colleges need to be given in the daytime, and this necessitates a provision for shutting out the light which will interfere with the experiment.

The light may be excluded from a room by tight-fitting shutters, or with curtains. It is very difficult to make shutters so tight that all light is excluded by them. It can be done much better and cheaper by having some frames made the size of the window frames, and covering them with what is known as enamelled cloth, such as is used in upholstery and carriage trimming. These should fit tight enough in their places to remain when placed.

The same kind of cloth can be attached to common curtain fixtures, and rolled up and down as wanted; but it will be found that a great deal of light will pass by the edge of these curtains. This can be obviated by tacking strips of the same material a foot wide to the side of the casing, so that the curtain will roll down inside of the strips. When sunlight or the lime-light is used, it is not always necessary that the room should be totally dark; and, indeed, some of the best experimenters think it a part of their success that their work is done in a room that is light enough for one to see to read a newspaper. Yet there are some experiments which require that extraneous light be shut out from

the room : for instance, the projection of the Fraunhofer lines in the spectrum of the sun, and the phenomena
of diffraction. For these, and the like, the darker the
room the better.

The curtain in the window that holds the *porte lumiere* will need to have a hole cut in it large enough to
allow the beam of light to come through, and to permit
the hand to give proper motions to the mirror. A flap
should hang over this when sunlight is not wanted, and
the electric light or the lime-light is used instead.

THE SCREEN.

The white surface that receives the projected picture
is called the screen, and it may be a white finished
wall, or white cloth properly mounted. The back of a
large wall-map makes a good screen if the light is used
in front of it, and only a small disk of light is needed,
but the backs of such maps are apt to get discolored,
and to become so dark as to be useless. They may
then be made white by painting them with whiting,
mixed in a thin solution of glue.

For a parlor exhibition, a common sheet may be
hung against the wall, or between the folding doors,
and the lantern used on either side. If the lantern is
placed back of the screen, the latter should be kept
wet, as it is made more translucent, and the pictures
will appear brighter.

When the *porte lumiere,* the electric light, or the oxyhydrogen lantern is used, a much larger screen will be
necessary. They are sometimes made twenty-five feet
square or more, but for most purposes a screen fifteen
feet square will be large enough. Common bleached
sheeting, ten quarters wide, can be bought in most
towns. A strip of this, ten yards long, cut into two

pieces of equal length, and having the selvedges sewed together, will make such a screen with but one seam. That these edges may come together, but not lap, let the sewing be done with what is called the carpet stitch. Some loops of tape or small rings may be sewed into the corners, and it may be hung upon nails driven into the wall at the proper places.

It is often convenient to have the screen so mounted as to permit it to be rolled up when not in use, and various devices have been invented to effect this. Perhaps the neatest is to have a roller at the top containing a strong spring, which is wound up when the screen is pulled down, — a large curtain fixture. A wooden roller sixteen feet long is likely to sag in the middle, unless it is made so large as to be cumbersome. It is best to have one made of tin tube about three inches in diameter.

A screen can quickly be put up in any room by procuring two strips of board, two or three inches broad, and long enough to reach from the floor to the ceiling. Fasten the sides of the screen to these, and then wedge them tightly between the floor and the ceiling. A portable frame which can be adjusted to various heights may be made by having two such strips for each side: one of them to be provided with a collar at its end for the other to slide through, and to be made fast together by a thumb-screw through the collar, as in the figure.

This will permit one to adjust it to different heights to its limit of eighteen or twenty feet, while by resting the foot upon chairs or tables a still higher room would be provided for.

CHAPTER II.

ARTIFICIAL LIGHTS.

While it is true that sun-light is much brighter than artificial light, and is therefore very desirable as the illuminating agent in projections, it is also true that sun-light is not always to be depended upon, and it will frequently disappoint one, by reason of clouds, which will entirely prevent using the *porte lumiere* and the experiment will need to be postponed until the sky is again clear. In some circumstances such delay would be no serious matter, and one could very well wait ; at other times the delay would be very inconvenient and work some harm in our educational institutions ; hence recourse is had to artificial light and lanterns. As nearly every kind of projection is possible in this way, and some persons will be provided with such instruments, and still others who would like to know what can be done with lanterns, some space will be given to descriptions of some of their more common forms and their applications.

THE ELECTRIC LIGHT.

Chief among the artificial lights used in projecting is the electric light, which is produced when a powerful current of electricity is made to pass between two carbon points which are separated a short distance from each other. It is necessary to have a current of elec-

tricity from forty or more Grove or Bunsen cells to produce this light, and there is then needed a special kind
of lamp furnished with some mechanism that will auto-
matically keep the two carbon points at a proper height
and a certain distance apart. Such lamps have been
devised by Dubosque of Paris, and Browning of Lon-
don, and others; but the best of them are not constant,
except with a very powerful battery, and when used
with only forty or fifty cells will need personal attention
every few minutes. Browning has advertised a small
electric lamp, which he says will give a constant light
with only six or eight cells. A number of these small
lamps have been brought to this country, but, so far as
the writer knows, no one has been able to work them
with anything like so small a battery.

There are several reasons why the electric light is
not more generally used for this and other purposes.
First, its cost: the battery with lamps costing about
two hundred dollars. Second, the consumption of zinc,
acids, and mercury for amalgamation, which, with the
labor of setting it up and cleaning it after use, may
be reckoned at ten dollars a day. Third, the noxious
fumes which constantly arise from a working battery,
making it necessary to have a special battery-room, well
ventilated; and fourth, the need of frequently over-
hauling it, re-amalgamating the zincs and filing the wire
connections. These have made every one who has ever
worked with a battery, wish that some substitute could
be found for it. The magneto-electric machines devised by Wilde, Ladd, Farmer, and others, have been
more or less successful, but have been much too costly,
and require eight to ten horse-power to run them.

The machine that promises the most for us now is
the one known as the Gramme machine, a French in-

vention. All who have seen its performance speak in high terms of praise of it. At present there is but one of them in the United States ; that one belongs to the University of Pennsylvania, at Philadelphia. Its cost was about $1,000. It needs but one or two horse-power to run it eight hundred revolutions a minute, when it gives a light equal to 1,600 candles. The latest pattern, made especially for produciug the electric light, weighs 400 pounds, is run by one-horse power, and gives a light equal to 2,000 wax candles. It would seem as though this was the thing we have so long waited for. It is now being used for lighting up large manufactories, only four lamps being needed in a room three hundred feet long, which is so well lighted as to leave no shadows.

MAGNESIUM LIGHT.

Wire made of the metal magnesium can be lighted like a piece of pine wood, and continues to burn with a most brilliant and dazzling light. In order to regulate the burning of the wire or narrow ribbon, which is generally employed, a lamp with a feed run by clock-work is used. Sometimes two ribbons are burned at the same time. This light is not constant, and is even more liable to go out than the electric light, and furthermore a special arrangement of cloth tubing is used for carrying away the product of the combustion, magnesium oxide, a bulky white powder, which accumulates very rapidly.

The cost of a lamp is about fifty dollars, and the cost of the magnesium is about two dollars an hour. It has the great advantage of being very compact, requiring but a few minutes to prepare at any time, and giving then a light which is amply sufficient for any exhibition, and is especially well adapted for experiments in fluorescence on account of the abundance of ultra violet rays. A three-inch transparency can be magnified up to thirty feet in diameter, and be well enough lighted for a large audience to see plainly.

THE OXYHYDROGEN, OR DRUMMOND LIGHT.

A very intense light is produced by projecting a blow-pipe flame of mixed hydrogen and oxygen gases upon a stick of unslacked lime. The great heat raises the lime to vivid incandescence. Sometimes magnesia is employed instead of lime; it is then called magnesia light, and when zirconia is used it is called zirconia light. The two latter are seldom used in the United States, but the former is very common. The gases are stored in portable tanks or india-rubber bags, which are connected by flexible tubes to the jet, from which it is driven by pressure upon the bags, and is lighted at the outlet. There are many patterns of these jets, some of them permitting the gases to mix before their escape, and others not until they are ignited. The mixed jet is the most economical one, and is to be preferred for most purposes. Such a piece of apparatus had better be bought of a reliable dealer.

Common illuminating gas can be used in place of pure hydrogen, but the light is not quite so intense. The demand for these gases has been so great during the past three or four years that they are now manufactured on a commercial scale in New York city, and are compressed into copper tanks holding from ten to sixty cubic feet. These tanks are exceedingly convenient. They retail for twenty-two cents a foot for oxygen, and eight cents a foot for the common gas.

An alcohol flame is sometimes used with oxygen, the jet supplying the latter forcing the flame upon the piece of lime. The alcohol is fed from a reservoir seen at the back of the lantern. This gives an excellent light,

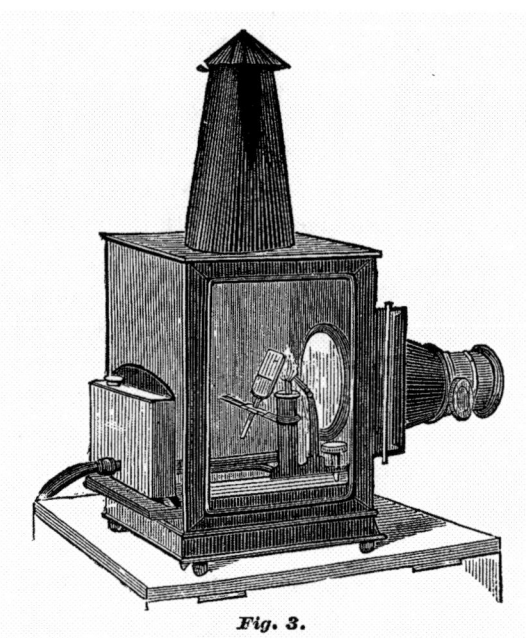

Fig. 3.

and quite sufficient for many purposes. A picture three inches in diameter may be magnified up to fifteen feet and be well lighted. The light produced in this way is called the Bude light.

OTHER LIGHTS.

Common illuminating gas may be employed with advantage where the room is small, and great intensity is not required. The form of burner known as the Argand, is best for this use.

Kerosene and lard oils have been, and are still

largely employed. When burned under the most favorable conditions, kerosene will yield a light equal to thirty or forty candles, and will illuminate a disk eight or ten feet in diameter very well.

Illuminating gas and kerosene have this advantage, that they are very cheap, costing but a few cents an hour ; but they can only be used to enlarge pictures which are very transparent, outline drawings, or chemical reactions in the large tank, which will be described further on.

There is no absolute standard of the luminous intensity of light. A conventional standard of its intensity is used for convenience, and for regulating the illuminating power of gas. In Massachusetts, the legal standard is the light of a sperm candle weighing two and two-thirds ounces, when consuming one hundred and twenty grains in an hour. In some places wax candles are adopted, but the light does not differ much in intensity from that given by sperm candles.

With such a standard as this, the various lights which have been specified, have the following relative illuminating powers under favorable circumstances :

Illuminating gas, ...	15
Coal oil in Argand burner, ...	20
Coal oil in Marcy's lamp, ...	25
Magnesium, ...	40
The lime light with oxygen and alcohol, . .	50
" " " " " " common gas,	100
" " " " " " hydrogen, .	125
The Electric light, . . . 500 to 10,000	

All of these are very variable, especially the lime light, which depends upon the quality and pressure of the gases, the quality of the stick of lime, and the kind of jet used.

The electric light has a very wide range in intensity,

as will be seen. It has not been found possible yet to make an electric light with less intensity than several hundred candle-power, while with the new gramme machine before mentioned, it is expected that not less than twenty thousand candle-power will be obtained.

Sun light has about four times the intensity of the most powerful electric light that has yet been measured. When it is reflected from a good mirror, it loses its brilliancy somewhat, but is second to nothing but the direct sun light. Hence, the desirability of employing the latter whenever it is possible, for both efficiency and cheapness.

LANTERNS.

When sun light is used, the room must be darkened, and the light only admitted through the opening of the *porte lumiere;* but when artificial lights are used, it becomes necessary, not only to have the room dark, but to inclose the light, so that what is not used shall not interfere with what is used. The refractive power of lenses is made available for the purpose of securing a larger amount of light than is possible without. This will be understood by the diagram. Let *a* be a point of light, and *b c* an object three inches long and distant six or eight inches from the light.

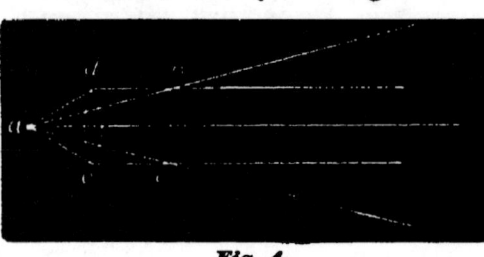

Fig. 4.

All the rays which will fall upon the object will be included in the angle *b a c.* Interpose the lens *d e* between the light and the object, and all the rays included in the much greater angle *d a e* will now fall upon *b c*, and it will be much brighter. There is always another lens, and sometimes two, called the *objective*, in front of the lantern, to give definition to the

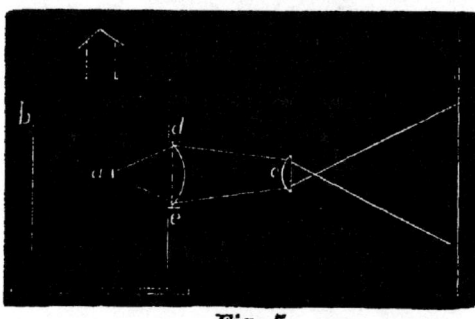

Fig. 5.

picture. All the essential parts of a lantern are shown in the accompanying picture : — *a* is the source of light inclosed in the box *b; d e*, the lens for diverting a greater number of light-rays, and called the *condenser; c* is the *objective;* and *s*, the screen to receive the light.

Fig. 6.

The forms of lanterns differ somewhat as they are
adapted to different purposes, and they are called stereopticons when the electric, the magnesium, or the
lime light is used in them.

The electric lamp is generally placed within the lantern box, which is made to accommodate either it, or
the lime light, as is most convenient or desirable.

The magnesium lantern shown in Fig. 6 (p. 15) has
the lamp set into it behind, and resting upon a floor
which can be elevated or depressed by a thumb-screw.
The flexible tubing is connected with the chimney, and
the magnesia is thus prevented from distributing itself
throughout the apartment.

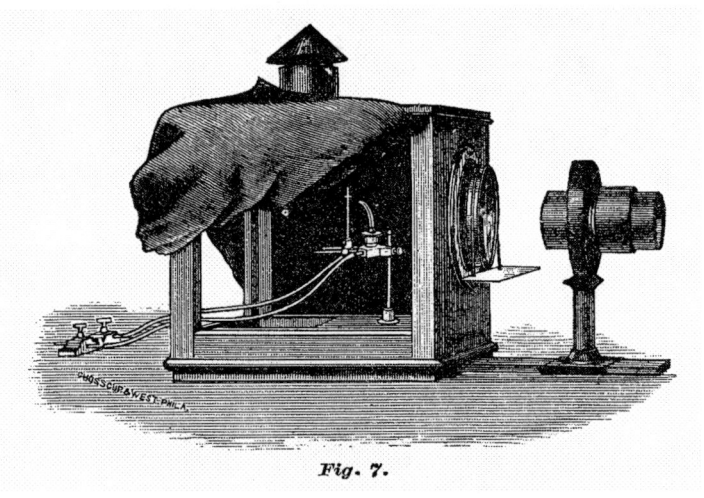

Fig. 7.

The above engraving (Fig. 7) represents the essential conditions for the lime light in a very convenient
form. Within the lantern may be seen the fixtures for
holding the lime and the jet, with the rubber tubing to
connect with the gas tanks. The front side and back

are provided with heavy black-cloth curtains which may easily be raised to adjust the fixtures within, yet shut

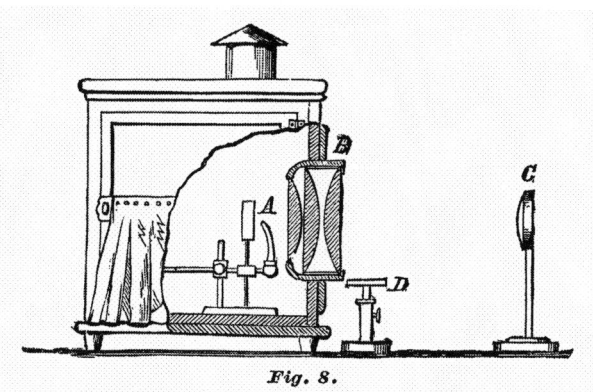

Fig. 8.

in the light when allowed to hang free. The condenser consists of three lenses, the front one being five inches in diameter. This is shown in the cut (Fig. 8)

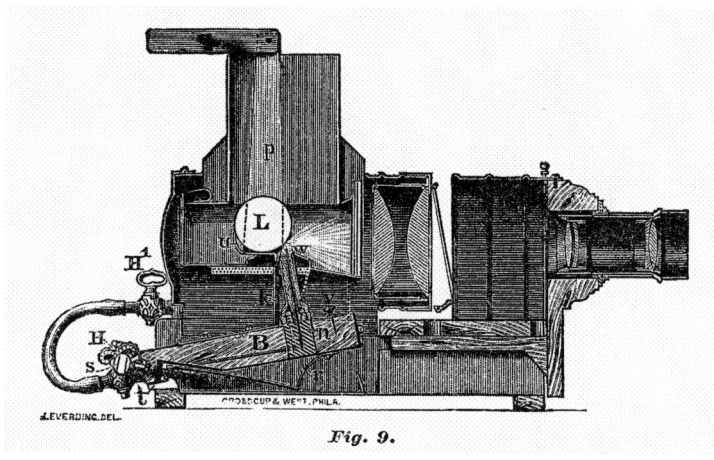

Fig. 9.

of the transverse section of it. The objective represented is also compound. This form of lantern is made by George Wale & Co. at the Stevens Institute, Hoboken, N. J., and called the Experimenter's Lantern, because of its adaptability to many kinds of experimental work, as well as to the exhibition of photographic transparencies.

Mr. L. J. Marcy, of Philadelphia, has, in an ingenious manner, made a lantern jet which can be used with alcohòl, common gas, or hydrogen, to produce the lime light. He calls it the triple jet. The engraving (Fig. 9) shows the lantern in section. L being the disk of lime, the gases ignited at w. This lantern is compact, light, and has a very convenient arrangement for holding slides, tanks, and so forth.

The oil lanterns have various names given to them by different makers, such as magic lanterns, lamposcopes, sciopticons, and so

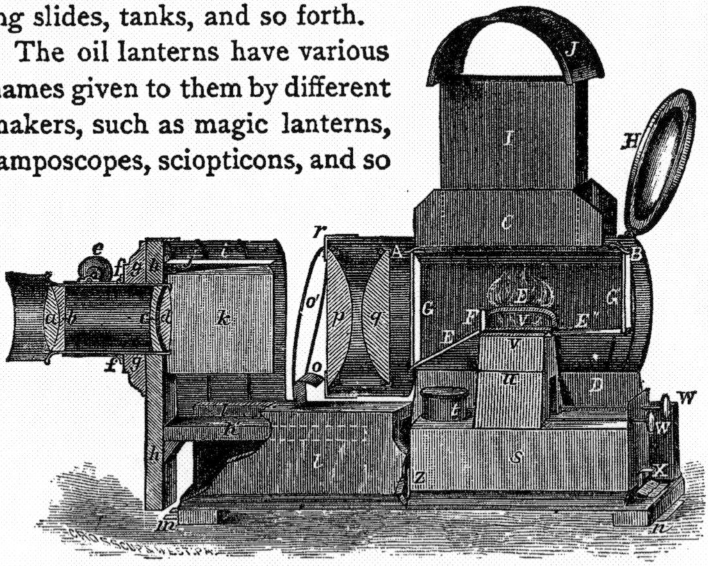

Fig. 10.

on. Fig. 10 represents the new form, the Sciopticon, which for simplicity, compactness, and brilliancy of illumination, surpasses every other oil lantern in the market. Neither the Sciopticon, nor any of the stereopticons, are difficult to manage. But little time will be needed to learn all that is needful to know in order to work them well, and they are not likely to become disarranged.

The course of experiments to be given will be generally adapted to both the *porte lumière*, and the lantern, but the adjustments will be described for the former. If, however, some special arrangement of the lantern will be needed for a given experiment, it will be pointed out.

A sufficient number and variety of experiments in physics, chemistry, and natural history, will be described, to make any one who is practically interested in the work so familiar with the apparatus and its working, that he will need no further instruction in the art of projecting.

LENSES.

All the lanterns in the market are furnished with the proper lenses for the ordinary kinds of projections, such as transparencies, etc. Solar microscopes are also generally well provided with lenses suitable for the work to be done with them, but as this treatise is for the use of those who have neither, yet who would like to experiment, some things, necessary to be known about lenses, and their uses are mentioned here, chiefly as to their adaptability to the experiments with the *porte lumière.* In Fig. 11 the rays of light are shown to fall upon the

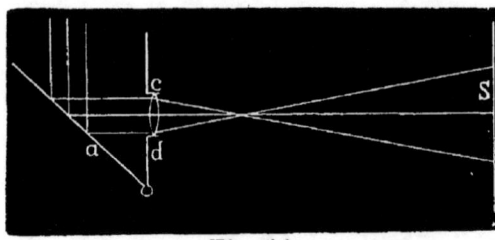

Fig. 11.

double convex lens *c d* when they are converged to a focus, and afterward they separate and fall upon the screen *s*. It may be noticed that this lens, used next to the orifice, is represented as a double convex lens in nearly every place in the book. Such a lens, when properly constructed, has less spherical aberration than any other form. To be thus properly constructed one side should have a greater convexity than the other, the radii of curvature for the two sides being as one to six, but such lenses are not enough better than the more common form of having the two sides of equal convexity, to make any difference except for special work, if the lens is used as a condenser: so that for this purpose a common double convex lens, with the faces of equal curvature, will be found to answer for most purposes. But a plano-convex lens with the same focal length as the convex lens is nearly as good.

Here it may be remarked that whenever a plano-convex lens is used it should have its convex side turned toward the rays which are parallel, or most nearly parallel. Thus the lens (Fig. 12) has its convex side toward *a*, where the rays are parallel, and should be so whether the rays

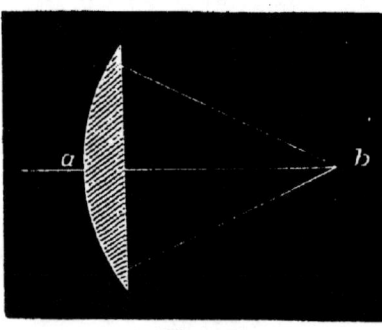

Fig. 12.

enter that side or emerge from it, when the source
of light is at *b*. It must not be inferred that nothing can be done except with lenses of a particular
sort. On the contrary very much can be done with
such poor lenses as are used in dark lanterns, and
are full of striæ and air bubbles. The lenses that
come with ordinary magic lanterns will answer for many
purposes. Spectacle glasses, linen provers, botanical

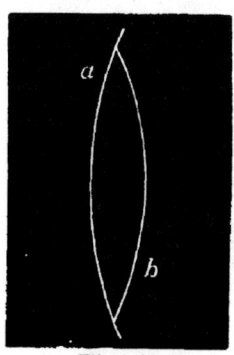

glasses, are all very useful. A powerful lens can be made out of two
watch glasses, one (*a*) a little larger
than the other (*b*). Bring the two
together under clear water. When
raised out of the water they will
adhere quite strongly, and for a time
can be used to advantage as a mag-
nifier.

Fig. 13.

FOCAL LENGTH.

The focal length of a lens should
be known before it is brought into use, and it may be
determined experimentally in the following way :

With the lens placed as shown in Fig. 12, so that the
parallel rays from the *porte lumiere* fall squarely upon
it, measure the distance from the centre of the lens to
the point *b*, the focus. If the lens be double-convex,
add one-half its thickness to the measured line. This
number will represent the focal length of the lens. If
the lens be plano-convex, its focal length will be the
distance from its flat side (Fig.12) to the focus. Again,
hold the lens so that the direct rays of the sun fall perpendicularly upon it, and measure as before the distance to the focus. Lastly, if the sun be not shining,
bring the lens close to a white wall or sheet of paper
opposite to a window, and hold it so that the light from

the window falls perpendicularly upon it; then slowly move the lens from the paper toward the window, until the inverted image of some object out of doors, such as a cloud, or house, or tree, half a mile or more away, appears plainest. Measure the distance from the lens to the paper, as in the other cases.

If diverging rays had been used instead of parallel rays, the focus would have been at some greater distance; and if converging rays, a less distance than that indicated for parallel rays, and these differing foci may be infinite in number; hence, the focal length of a lens is always specified for parallel rays.

CARE OF LENSES.

Before using lenses, or other pieces of glass apparatus, see to it that they are clean, and, as far as possible, avoid touching them upon their polished surfaces with the fingers, as the latter will always leave a mark upon such a surface. A piece of cotton-flannel, or old fluffy linen will do to wipe glasses with. Wet the cloth with water, or, still better, with alcohol, if there are spots that will not otherwise come off. One may know when a piece of glass is clean by gently breathing upon it, and then noticing how long the condensed moisture remains upon its surface. If it is really clean, the moisture will disappear in a second or two; if it remains for eight or ten seconds or longer, the glass is not clean, though it may appear to be so to the eye. Be careful to keep such pieces from touching anything harder than the cloth they are wiped with, for even wood will scratch a nicely polished glass surface. If these pieces are not mounted in such a way as to protect them, they should be laid, when not in use, upon a piece of cotton-flannel or velvet.

MOUNTINGS FOR LENSES.

Lenses may be purchased already mounted in any desirable way, but generally the mounting costs as much or more than the lens. What is needed is a fixture that will hold the lens at a proper height, and has a considerable latitude of movement in every direction. For many purposes an unmounted lens can be held in the pinch of a common retort-holder as in Fig. 14.

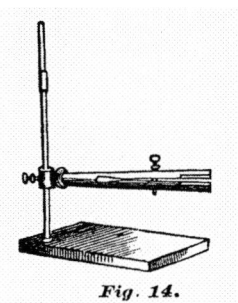

Fig. 14.

It is often convenient, and sometimes necessary, to have the lens so mounted as to cut off light that would otherwise pass by its edge and mar the effect upon the screen. To do this it is only needful to cut a hole the size of the lens, in a square piece of board (Fig. 15) having the proper size and thickness, and fasten the lens in it with small brads or triangular p i e c e s of tin or zinc, such as glaziers use for fastening window-glass in the sash. The board should be thicker than the lens, and the latter should be so sunk into it that its surface will

Fig. 15.

not touch the table when lying upon it. This will prevent it from getting scratched. Thus mounted, the board may be held in the retort-holder as before.

The smaller lenses, having a focal length of not more than an inch or two, had better be bought already mounted for carrying in the pocket, such as botanical glasses and linen provers, or, if it can be afforded, get one of Zentmayer's gas microscope objectives.

For holding pictures, and other objects, for projection, the retort-holder will answer, in many cases, just as well as a more costly fixture. Not only may the larger pieces, such as photographic transparencies, leaves of trees, and the like, be held well in it, but microscopic specimens in glass slides, also small unmounted objects, such as parts of flowers, insects, etc.,

Fig. 16.

may be held in small forceps (Fig. 16), which in turn may be held in the retort-holder. It will be found convenient to have as many as three of the latter.

PROJECTIONS.

TO PROJECT WITH THE PORTE LUMIERE AND A SINGLE LENS.

Fasten the *porte lumiere* in its place, and so adjust it that the beam of light *l* (Fig. 17) is reflected horizontally, and falls upon the screen *s*. It will appear as a bright spot, five or six inches in diameter. Darken the room, by drawing the curtains or closing the shutters, and the beam of light can then

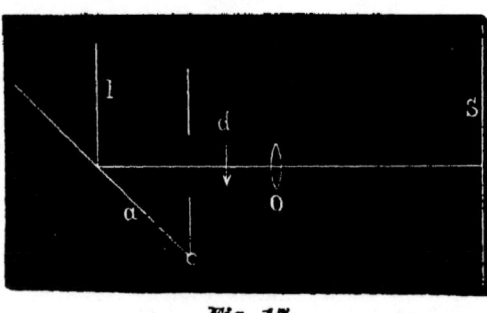

Fig. 17.

be seen from the window to the screen by the light re-flected from the dust particles, which are always in the air. Now fasten in the retort-stand a lens *o* four or five inches in diameter, and with a focus of a foot or more, and place it two or three feet from the opening, in the path of the beam, and perpendicular to it. It will at once be noticed that the light is converged by the lens, the rays crossing each other in front of it, at its focus, from which they diverge, and appear upon the screen as a large disk of light. If some object, as *d*, (Fig. 17), be placed between the opening and the lens *o*, a place may be found by trial, when the image of the object will be seen upon the screen. The outline should be well defined ; it will be inverted and much enlarged.

Finding the right adjustment of the object and the lens, so that the image is in its proper place, and has a sharp outline, is called *focusing*. In general it is best done by fixing the object in the path of the beam first, and then placing the lens rather close to it, and slowly moving the lens toward the screen, being careful to keep it perpendicular to the beam until the image is plainest. It will be well for a beginner to take a number of ob-jects : some opaque, like the finger, a pencil, or a key, and some transparent, as a grasshopper's wing, or a piece of glass with a design drawn upon it, or a regu-lar lantern transparency. A lens thus used to project a picture is called an *objective*.

These two pieces of apparatus, the *porte lumiere* and the single lens, have a much wider application than one unfamiliar with them might suppose. Every picture made for the magic lantern, or the stereopticon, can be shown with these in the day-time, even better than with the others at night. Every school in the land may have one, for the carpenter can make the *porte lumiere*, and the lens will cost but a trifle. The pictures themselves, though not half as costly as they were before photography was applied in making them, can be rented of any one who keeps them for sale, if one cannot afford to buy them outright. Most excellent transparencies, on all sorts of subjects, can be bought, from six to nine dollars a dozen, of any lantern-maker or dealer in photographic materials.

If the teacher wished to give a lesson on the elements of drawing, his copies could be prepared upon glass, by one of the methods given a little further on. These, when projected, would be so large that a large school could see them as plainly as if they had been drawn upon a huge blackboard, with chalk. The room could be light enough for any of the required work. Geometrical figures, outline maps, botanical specimens, the kaleidoscope, chemical reactions in a large test tube ; natural history specimens, such as small fish, pollywogs, water beetles, butterflies, grasshoppers ; the splendid colors on huge soap-bubbles ; the vibrations of tuning-forks, and of cords ; the intensity of light, reflection, refraction, magnifying power of lenses, and many more things, may be projected, in an admirable way, with only these two pieces.

THE CONDENSER AND ITS USE.

The rays of light reflected from the mirror *a* (Fig. **17**) through the aperture, are parallel, and the diameter of the lens *o* should be as great as the thickness of the beam, and it may have a greater diameter with advantage. If it has less, some of the light will pass its edge, and either be unused or, what would be worse, fall upon the screen and make a bright spot in the middle of the picture. The smaller the object to be pro-

jected, the smaller must be the lens used as the objective, and the shorter must be its focal length ; hence, if
a beam of parallel rays is used, it may often be so small
as to be nearly useless, for the divergence is so rapid
beyond the focus of a short focus lens that the little
light thus used would be too much scattered. It becomes necessary, then, to make a large beam of light
to pass through the small lens. This is accomplished
by means of a second lens, called the *condenser*, because
its office is to condense a large number of light rays for
the double purpose of illuminating the object better
and making them all to pass through the smaller lens.
This condenser is usually four or five inches in diameter,
though for special purposes it is sometimes made a foot
or more in diameter. For the *porte lumiere* the condenser may be the same lens that was used as the objective, or any lens may be used that has a sufficient diameter and a focus of from one to two feet. It may be
either double-convex or plano-convex.

TO PROJECT WITH A CONDENSER.

The object *d* (Fig. 18) is placed near the condenser
c, and the objective *o* is brought near to it and slowly

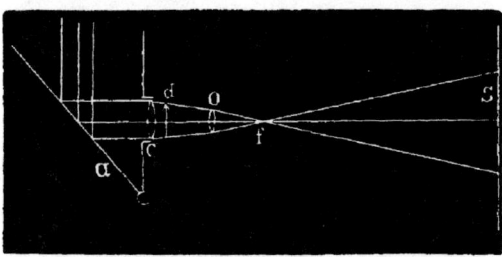

moved toward
the screen, as
before, until
the well de-
fined image ap-
pears upon it.
It must be not-
ed here that

Fig. 18.

the size and focal length of the objective must be such
that all the light passes through it when it is at its
proper distance from the object. If *o* be moved toward

the object, it will be seen that some of the light does not pass through it. If the object *d* be moved toward the objective, then some parts of it will not be lighted, and consequently but a part of it would be projected. If the object *d* is quite small, like a fly, or a flea, or a small crystal, it will be necessary to bring it forward, toward the focus of the condenser, where it will be more strongly lighted, and allow the use of an objective of shorter focus, and consequently higher magnifying power. If the object be made of wood, or any kind of tissue, be careful about bringing it very near to the focus, as the great heat there may destroy it in a few seconds. This danger may be somewhat lessened by placing between the condenser and the object the chemical tank, containing a strong solution of alum. The common pocket botanical glass, having a focus of an inch or two, will answer for very much of this work, but Zentmayer's inch-and-a-half gas microscope object-ive is superior to any other lens I have seen for such projections.

This arrangement is essentially the *solar microscope.* The object may be exceedingly minute if the objective has a very short focus, say half an inch, or less.

It is possible to magnify an object a thousand diam-eters, or a million times, and still have it so well lighted that a large audience can see it plainly. A list of things that are suitable for projections with this ar-rangement is appended, mainly for the purpose of indi-cating the breadth of its field of usefulness :

Hairs of various animals, which may be held between two strips of glass. Down from the wings of moths and butterflies ; these will adhere to a piece of glass with-out any pressure. Scales of fishes. Eyes, legs, and wings of flies, or the whole of any insect. Stings of

bees and wasps. Antennæ of moths and mosquitoes.
Fibres of cotton, woolen, silk, etc. Ferns, moss, lichens,
leaves of trees. Thin sections of wood. Small flowers,
stamens and pistils, pollen. Mites in cheese. Butterflies, beetles, animalcules in stagnant water. Vinegar
eels. Crystallization of camphor, sulphate of copper,
and most solutions of crystallizable substances.

Diatoms, and indeed most objects that are prepared
for the microscope, appear to good advantage upon a
screen. Any book upon the microscope will have
many valuable hints upon obtaining and preparing objects in a suitable way, and will be a very useful book
to one interested in natural history but who cannot afford to buy a good microscope.

OUTLINE DRAWINGS FOR PROJECTION.

Every one who uses either a lantern or the *porte
lumiere* for purposes of instruction, will need to make
outline pictures to illustrate his subject, as it will be frequently impracticable to get a photograph of what is
wanted. Moreover, a simple outline is often quite sufficient for the illustration, as, for instance, superposi-
tion and inclination of strata in geology ; sections of
machines ; writing, or musical notes ; outlines of leaves,
roots, parts of a flower, insects, maps, etc.

The surface of transparent glass is so smooth that it
cannot be marked with either common ink or a lead
pencil. If the glass be ground, so that a pencil will
mark it, it becomes so opaque that but little light can
go through it ; hence, a surface must be prepared which
will be transparent and yet allow marking upon it.
This can be effected in many ways, and I give a number which I know to be practicable :

1. If a piece of glass be rubbed on one surface with

a piece of hard soap, enough will adhere to it to make the glass semi-opaque. Now draw the design with a fine-pointed stick. It will clear the soap from the glass, and so permit the light to shine through the marks. This has the advantage of permitting the same glass to be used like a slate, for with a drop of water upon the finger the old design can be rubbed out, leaving the glass coated for another picture. The same thing can be done with a surface of beeswax, but the glass would need heating in order to re-spread the wax.

2. For more permanent pictures, a very good way is to flow the glass with photographers' transparent varnish, and then scratch the design upon the varnish, *not* cutting through to the glass. The light is so much scattered from this scratched surface, that it appears as a dark line, and answers a very good purpose. The prepared plate can be laid over the design wanted if it is to be a copy, and is of proper size ; the transparency allows the picture to be plainly seen, and all its markings can easily be followed. The varnish is quite hard when dry, and with a little care in handling these pictures, they need not become scratched. They can be entirely protected from that danger by covering them with another clean glass of the same size, and binding their edges with paper, as common lantern-pictures are bound. Photographers have also another kind of varnish called ground-glass varnish, which, when spread upon glass, gives it an appearance similar to ground glass. This surface permits drawing with a pencil or with ink upon it, and then a coat of the transparent varnish will render the first coat transparent, leaving the lines in ink or pencil ; or the design may be drawn *through* the first coat of varnish, in which case, the light will shine through the lines and appear white upon the screen.

3. If india ink be rubbed up in water until it is quite thick, it can be used for drawing designs upon ordinary glass.

4. A thin sheet of gelatin may be treated like the glass coated with the transparent varnish, and either have the design scratched upon it, or drawn with ink. It should be inclosed between two glasses for protection.

5. Thin sheets of transparent mica will receive lines drawn with india ink, or the figures may be scratched upon them with the needle or awl.

6. Designs may be nicely etched upon glass, by first coating the glass with a thin, even coat of beeswax, which can be well done by heating the glass over a lamp until the wax melts and flows over its upper surface. When it is cool, draw the design with a needle point or a small awl, cutting through the wax all the way. Take an old saucer, or some such dish which you are willing to spoil for other use, and put into it a table spoonfull of powdered fluor spar. Upon that pour a table spoonful of strong sulphuric acid, and stir them together with a stick. Fasten the glass, drawing uppermost, to a piece of board large enough to completely cover the dish. The fastening can be done by crowding tacks into the wood, so that the heads shall lap the glass and keep it in its place. When thus fixed and laid over the mixture of spar and acid, gently heat the dish, being careful not to inhale the fumes that will escape. When the fumes begin to appear, put the whole either out of doors or in a good chimney draught, and let it remain eight or ten minutes, when the wax may be removed by heat and rubbing, and the drawing will be found etched into the glass.

Beautiful pictures of crystals can be made in this way, by taking various crystallizable salts, such as am-

monium chloride, cupric sulphate, etc., and making a
rather dilute solution of them, and then adding a little
dissolved gum arabic. Flow the solution over the
plate, and let it remain horizontal until it is dry. The
crystals will be seen to have separated from the gum,
which will fill up all the intermediate space. Put over
the etching dish as before. The crystals will quickly
dissolve, and their outlines will be beautifully etched
upon the glass, which may now be washed clean in
water.

7. Engravings may be transferred to glass by first
coating the glass with dammar varnish, or with Canada
balsam, and letting it dry until it is very sticky, which
will take half a day or more. The picture to be transferred should be well soaked in soft water, and carefully laid upon the prepared glass, and pressed upon it,
so that no air bubbles or drops of water are seen un-
derneath. This should dry a whole day before it is
touched; then with the wetted finger, begin to rub off
the paper at the back. If this be skillfully done,
almost the whole of the paper can be removed, leaving
simply the ink upon the varnish. When the paper has
been removed, another coat of varnish will serve to
make the whole more transparent.

8. A piece of glass may be smoked in the ordinary
way, and a design marked upon it. This makes a very
good and plain picture. If the design is needed for
keeping, heat some alcohol in a cup or small porcelain
dish, and hold the smoked side of the glass in the alcohol vapor for a minute or two, and afterward it may be
varnished with photographers' varnish, carefully flowing
it over the plate in the same way that plates are flowed
for photographic purposes.

TO DETERMINE THE MAGNIFYING POWER OF A LENS, OR A COMBINATION OF LENSES, IN PROJECTING.

It will be evident, upon inspection, that the farther away the screen is from the lens, the larger will be the picture ; but for a given projection, the simplest way of determining the magnification is to choose some object of known dimensions for projection, and then to measure its size upon the screen. Suppose it be a lead pencil having a diameter of one fourth of an inch. If its image is a foot in diameter, it is evident that it is magnified $4 \times 12 = 48$ diameters. If it is three feet in diameter, then it has been magnified $4 \times 12 \times 3 = 144$ diameters. It will be convenient to have a scale, either photographed or etched upon glass, for the purpose of directly showing the magnifying power of lenses.

A *vernier* made upon glass by either of the described methods, will be convenient for study, and some measurements.

THE ANIMALCULE CAGE.

If one would exhibit the minute forms of life to be seen in water, an animalcule cage will be needed. This may be made in the following way:

Take two quite clear pieces of white glass, about four inches long and one inch wide. Two other pieces of

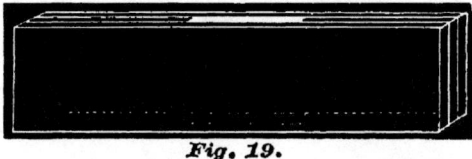

Fig. 19.

the same width, and one inch and a half long. Put these two shorter pieces between the longer ones, so as to separate them, and leave a space in the middle clear through. Fasten these together with japan varnish, being careful not to get any of the varnish into the opening. If any should get in, wipe it carefully out. When the varnish is dry, and the pieces are firmly fixed together, putty up the bottom of this hole so that it will hold water. When this is dry, it can be used to hold fluids of most kinds, but it is especially fitted for water containing animalcules, or vinegar with eels. It should be put back of the focus of the condenser, for the great heat there will boil the water in a little while, and the temperature of no more than 130° Fah. will quickly kill most all kinds of infusoria. Suitable water for examination can be found in old rain barrels, stagnant pools, water in which flowers have been standing for a day or two, an infusion of hay in water, and will be found very interesting. The larva of the mosquito is a lively and amusing thing when magnified to five or six feet in length.

THE CHEMICAL TANK.

For chemical experiments, and a variety of others, a tank of larger proportions will be necessary. The accompanying diagram (Fig. 20), shows the construction. Provide two pieces of clear, white glass, of the same size, about five inches by six, for the sides. These may be kept apart by a strip of rubber, about one-half of an inch thick, bent and cut at the corners, the whole clamped together by three or four clamps, as shown. If rubber with flat sides is not easily procurable, a piece of rubber tubing will answer

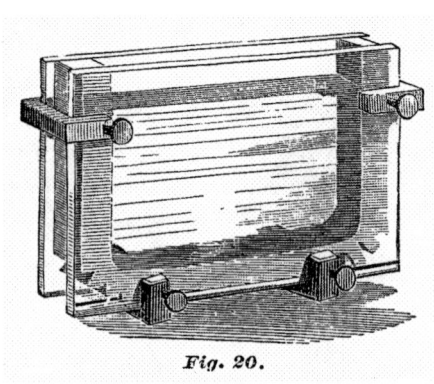

Fig. 20.

nearly as well; the tubing may be filled with sand to keep it firm. Such a tank will hold any kind of a solution, and may be quickly taken apart and cleaned. A tank which will answer for many experiments nearly as well can be made by cutting a semi-circular piece out of a board, of the proper size, and fastening the glass sides to it with cement. What is known as *marine glue* will be the best for this purpose, and as it is very convenient to have some of this glue for making and mending apparatus, because it will adhere to any surface, the method of preparing it is given : Dissolve, separately, equal parts of shellac and india rubber in naptha, and afterwards mix the solutions thoroughly, applying heat. It may be made thinner by adding more naptha. It may be preserved in a tin box. In order to use it, it must be heated, as well as the surfaces which are to receive it. Marine glue may be dissolved in ether, or a solution of potash.

A METHOD FOR PROJECTING LARGE PIECES OF APPARATUS.

Many pieces of apparatus used in illustration and demonstration are much too large to be projected in the ordinary way, as it is obvious that the size of the lens used as condenser will be the limit to the size of the object that can be shown with it. Thus, if sunlight is used, the diameter of the orifice c, d (Fig. 11), will be the measure of the largest picture that can be shown at once ; and if a lantern is employed, no picture larger than the condenser can be projected.

Suppose that it is desirable to show to an audience a piece of apparatus much too large for ordinary projection, and yet too small to be plainly seen, such, for instance, as the electroscope ; or the movement of a pith-ball under electrical excitement ; or the movement of a vibrating cord, or large tuning-fork ; or the apparatus for showing the linear expansion of metallic rods, etc. The following method will be found applicable to a great many such cases, where simply the outline of the instrument is needed.

Place a short focus objective (and the shorter the better), so near the focus of the condenser that all the light falls upon it. After refraction the light will form a very divergent beam and the focus in front of *o* will

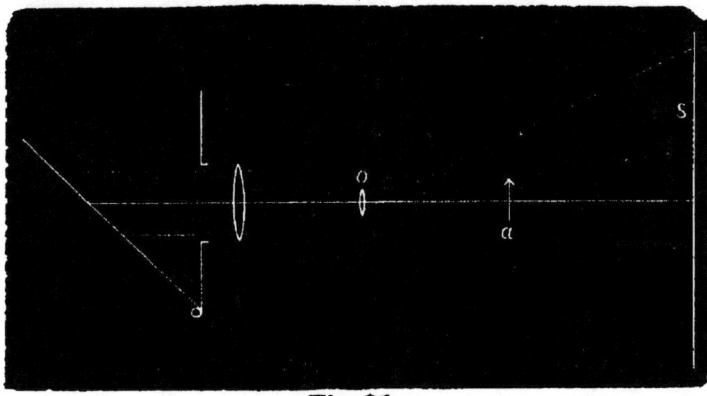

Fig. 21.

be a sharp point, practically a luminous point, and any object held between it and the screen *s*, will have a strong shadow cast upon the latter. The magnitude of this shadow will depend upon the distance from the focus. There will be no penumbra—the outline will be sharply defined.

If one has a lantern, the condensing lens above will answer without the objective, as its focus for parallel rays will be sufficiently short. A globular glass flask, filled with water and placed in the path of the rays, will also be found to be satisfactory. When a lantern is used instead of sunlight, it will be necessary to use the microscope attachment, which is described further on, working in *front* of the lens, the same as with the *porte lumiere.*

The following is a list of apparatus and of experiments which are suitable for such projection : Equilibrium of the same liquid in several communicating vessels; equilibrium of different liquids in communicating vessels; cartesian diver; the hydrometer; capillarity; diffusion of gases; Torricelli's experiment; Mariotte's law; the manometer; Sprengel's air pump; fountain in vacuo; the siphon; the pyrometer; the influence of pressure upon the boiling point; M. Despretz's experiment on the conductivity of solids; convection; the thermo-pile; umbra and penumbra; action of magnets; attraction and repulsion from electrical excitation. Natural history specimens, such as birds, rats, mice, squirrels, frogs, toads, live fishes, if in a tank with transparent sides; leaves of trees, ferns, etc.; well-defined crystals, such as quartz, feldspar, mica, pyrite; diagrams on glass of machinery, as the steam engine—these diagrams can be drawn a foot square or more; silhouettes, etc., etc., are all available with this method.

There is an advantage in this plan, when it is at all applicable, that will commend itself to every one, namely, it is available at any point between the focus and the screen, hence it will only be necessary to place the object in the path of the rays to the screen at such a point as will be convenient and will make the shadow sufficiently large. The instructor can stand by the object, and with a pointer like a pencil call attention to any particular part. And again, the field is so large that several objects can be in it at a time, if need be, for comparison, such for instance as leaves of several species of oaks or maples, or a range of capillary tubes of various diameters.

THE MEGASCOPE.

Photographs that are taken especially for projection with the magic lantern are often called transparencies because all of the lighter parts of the pictures are made as transparent as possible, and they are shown by light that is transmitted through them. If one would exhibit a picture like a stereoscopic view or a common *carte de visite,* it is evident that recourse must be had to some other arrangement. The light must be reflected from the picture, but when only the ordinary amount which is reflected from a surface of nine square inches is distributed over seventy-five or a hundred square feet, it is evident that it will be but dimly visible. If a large amount of light is concentrated upon the picture it will, of course, reflect more, and its image will be correspondingly brighter. This can be effected in two ways: first, by using a large lens, or second, by using a large concave mirror.

The following figures will serve to show how this may

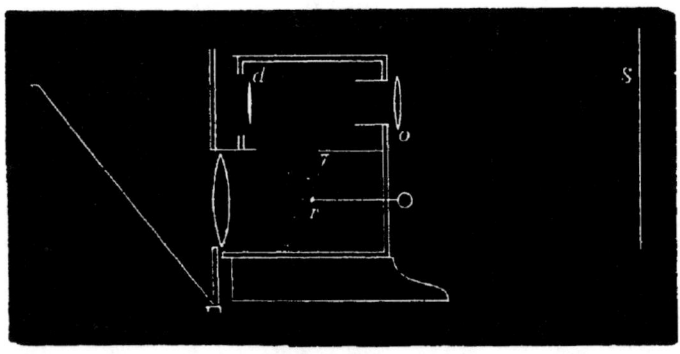

Fig. 22.

be done. When sunlight is used, the larger the condenser the better. One seven or eight inches in diameter, if possible, should concentrate the light upon a second plain mirror at *r*, which should have such an inclination as to reflect the converging rays upon the object to be shown at *d*, and strongly illuminate it ; the objective at *o* will be used in the same way as for any other projection. This apparatus should be in a box made with sides a foot square and six or eight inches deep. At the back of it a hole should be left at *d*, in which the various objects for exhibition may be held.

In place of the condenser and the plain mirror, a large concave reflector, such as is used behind lamps, may be placed at *r*, and the parallel rays from the *porte lumiere* allowed to fall upon it. It should be placed at such a distance from the object *d*, that it will just illuminate it ; this will of course be determined by the focal length of the mirror.

The room needs to be quite dark for the successful working of this apparatus, and especial care should be taken to prevent any of the light from the *porte lumiere* from being scattered into the room ; paint the box black, inside and out, with lampblack mixed in japan varnish.

If the lime light be used, as it generally is for such an exhibition, it is necessary to modify the lantern very much,—so much so as to require an entirely new instrument. The following is the simplest plan of one: A square wooden box made eighteen or twenty inches on a side, and about fifteen inches deep, may have a little way made in it on one side for the fixtures holding the jet *i* and the lime *j* to slide upon. A hole *r* cut six inches square, may be made near the corner, and another one on the front side for the light to come through upon the lens *o*, which is the only lens needed for work. The size of this hole should be no greater than that of the lens *o* used for the projections, but this lens should be as large as possible. A lens six or eight inches in diameter, with a focus of from eighteen to twenty-four inches, will be found best for the purpose.

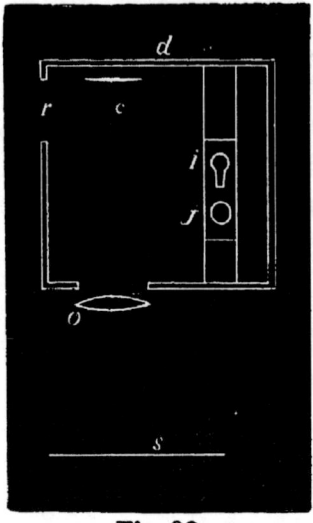

Fig. 23.

This may be held in the retort-holder before mentioned, and set at such a distance in front of the hole that an object *c*, when strongly lighted, will be plainly projected upon the screen *s*. The whole of the back on the in side should be covered with white paper. Let a black cloth flap hang over the hole at *r*, so that no light will enter the room, save what is reflected from the illuminated object.

With these conditions a dark photograph of an individual, upon a white background, will show quite well. Objects held in the hand, such as a watch with its movements, cameo pins, small flowers, surface of half an apple or orange. The latter, if squeezed when being shown, presents a very amusing appearance. Minerals, crystals, shells, bright-colored beetles, bugs, butterflies, etc., may all be exhibited, and appear, with the shades and shadows, like real objects. This constitutes the *megascope.*

The accompanying cut (Fig. 24) represents the scenic effect of the human hand, as projected by the megascope.

THE VERTICAL ATTACHMENT.

It is often very desirable to project such phenomena as the ripples upon the surface of water, the move-

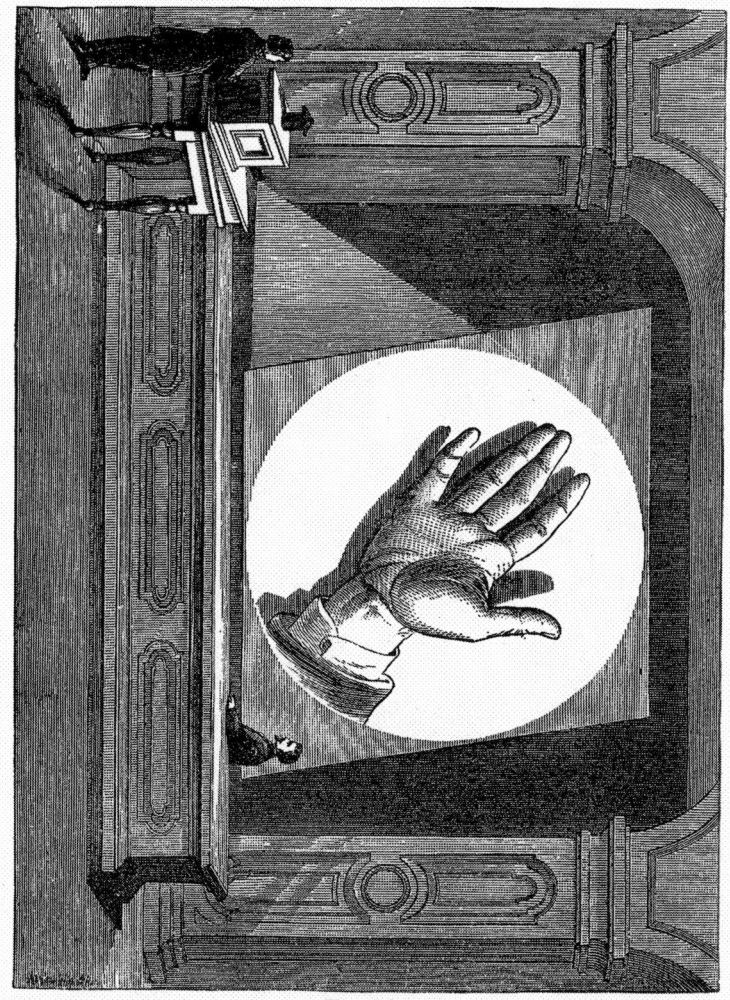

Fig. 24.

ments of a horizontal galvanometer needle, etc., such as cannot be exhibited with the common forms of apparatus for projections. At first the awkward method was adopted of turning the lantern up so that it rested upon its back. This endangered the condensing lenses of the lantern from the great heat immediately under t h e m. Professor Cook and Doctor Morton have greatly improved upon

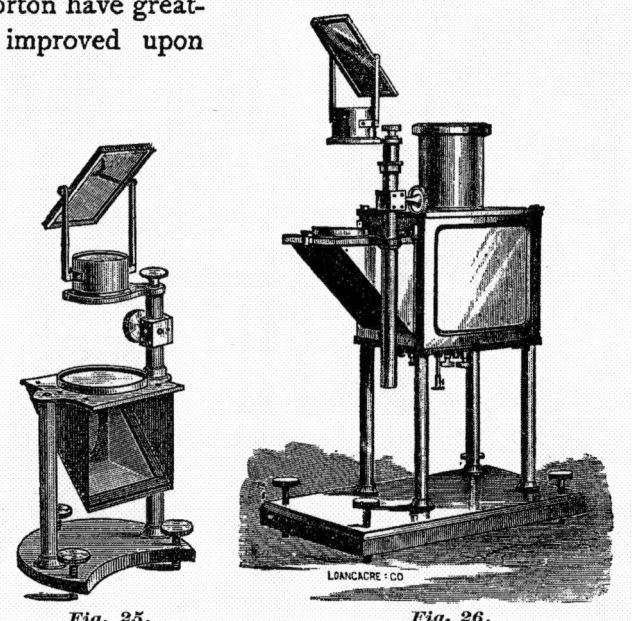

Fig. 25.   *Fig. 26.*

this, and have added a most valuable attachment to the lantern.

The cut (Fig. 25) represents this invention. It consists of a plane mirror inclined at an angle of 45°, and when so placed that the beam of light from the lantern falls upon it, it is reflected perpendicularly upwards upon a lens that converges the light when it passes through the objective above it, and falls upon a second mirror, which is so mounted as to allow reflection in any direction. The same device is made a part of the standard lantern of the country, and called the "*College Lantern,*" manufactured by George Wale & Co., of Hoboken, N. J. By an ingenious arrangement the change from the horizontal to the vertical can be made in less than half a minute. The microscope, the polariscope, the electric-light regulator, and several other fixtures, are fitted to this instrument, making it a most perfect and complete lantern.

Such a vertical attachment as is shown in Fig. 25 is applicable to the *porte lumiere,* but one can be extemporized, that will do good service, with such material as is accessible to every one. An iron filter-stand, such

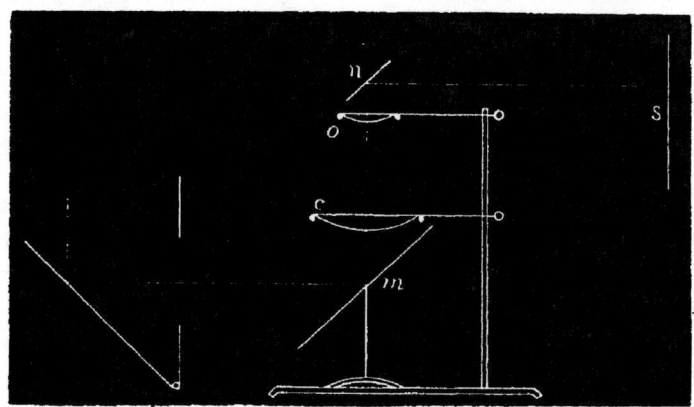

Fig. 27.

as is in common use in every chemical laboratory, may be taken, and the condensing lens *c* laid upon the lower or largest ring, and the objective, *o,* upon the upper or smaller one, as shown in Fig. 27. Below the lower ring a plain mirror *m* may be placed, at such an inclination that the beam of parallel rays falling upon it from the *porte lumiere* will be reflected upward through the two lenses upon another smaller mirror, *n*, which may be held in a retort-stand, and the beam directed to the proper place.

PHYSICAL EXPERIMENTS.

DIVISIBILITY OF MATTER.

A good way to show the minute divisibility of matter is to dissolve, in water, a small quantity, say a gram, of cupric sulphate, and add enough ammonia-water to make a clear, blue solution. Put it into the chemical tank, having measured its capacity in cubic centimeters, or inches, fill it with water, and project the tank by the method described on page 183. A beautiful blue color will appear upon the screen. With a small syphon of bent-glass tube, draw out one-half of the solution and fill up with pure water. The amount of coloring matter will be reduced one-half, but the solution will be strongly colored. Remove, in the same way, another half, and so on until the blue color is no longer visible —comparing the color with that of pure water, projected, at the same time, in a test-tube. Keep account of the number of dilutions, and at last, when the blue color is on the vanishing point, calculate the weight of cupric sulphate in each cubic centimeter of water. In place of the copper solution, any of the analine dyes will do as well.

The same thing can be illustrated with a soap-bubble,

blown thin, and projected in the diverging beam (Fig. 21). The bubble will be sharply defined upon the screen, and its magnitude will depend upon the divergence of the beam of light, and its distance from the screen. It may be made ten or fifteen feet in diameter, if the lens have a short focus. The colors will begin to appear around the pipe in bands, and computation of the thickness may be made, and of the probable number of molecules in its thickness. For the consideration of this, see "The New Chemistry," by Professor Cooke, and *Nature*, Vol. I, p. 551; also Galloway's "First Steps in Chemistry," article 102.

POROSITY.

The gases dissolved in common water will be expelled by gently heating some in a test-tube while the whole is projected. The bubbles will be seen to form and rise where nothing was before visible. The porosity of water can be shown by projecting a test-tube half filled with it, and its depth marked by a bit of thread tied about the tube at the level of the surface. A considerable quantity of salt or sugar can be added to the water without noticeably increasing its bulk. A piece of chalk dropped into a test-tube containing warm water will at once give out quite a quantity of included air.

The ordinary experiment of showing the porosity of leather by forcing mercury through it by atmospheric pressure into a partial vacuum, can be exhibited by projecting the upper part of the tube, while the exhaustion is going on. The mercury will be seen to fall upward on account of the inverting by the lens.

A mixture of equal parts of strong sulphuric acid and water loses notably in volume when cool. Fill a

test-tube with the fresh mixture, and tie a string about
the tube at the hight of the mixture. It will be too
hot for handling with the fingers at first, but it may be
cooled in a few minutes enough to show the shrinkage,
by stirring it in a dish of cold water. The surface will
be seen to be considerably below the string which
marked its original hight. This experiment may be
used to exhibit compressibility of liquids.

Most of the experiments which are suitable for projection of the properties of matter are chemical, and
will be found described under that head. Diagrams,
such as are given in most text-books on mechanics, can
be made upon glass by one of the processes described
on page 184, and will be found very convenient to a lec-
turer upon that subject.

COHESION.

A drop of water or other fluid exhibits this, and may
be projected with the lantern, or with the *porte lumiere*,
and a single lens (Fig. 28). Sprinkle a little lamp-

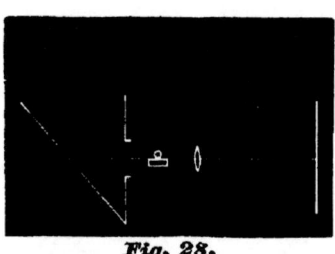

black or lycopodium-powder
upon one side of a strip of
glass, like a microscope slide-
and place it in the proper
place for projecting, keeping it horizontal that the
dust may not s l i d e off.

Fig. 28.

Now place a single drop of
water upon the slide; the powder will prevent it from
spreading upon the glass, and it will gather itself up
into a round globule with some of the dust over its surface, making an interesting object upon the screen.

Again, a saturated solution of zinc-sulphate is put into
a white glass square bottle, two inches square, and

three or four inches high. Let the bottle be about half filled with this solution. Into a few drops of bisulphide of carbon drop a piece of iodine. It will at once stain the bisulphide a dark-brown color, which should then be carefully dropped upon the solution of zinc, where it will float. If now pure water be carefully added, so as to rest upon the solution of zinc, the bisulphide will collect into an oblate spheroid, having the appearance of brown-colored glass. A square bottle will enable one to project it better, as a round bottle would make a cylindrical lens, and the projection would be indistinct, unless the vessel was quite large.

Nearly fill the large tank (Fig. 20) with alcohol, and project the tank with the lantern, or with the single lens and *porte lumiere.* Now drop upon the alcohol, with a glass rod, or other convenient thing, any of the aniline dyes. As soon as the dye touches the alcohol it will go straight down for a short distance, then it will branch, and these will shortly branch again, and so on to the bottom of the tank, when there will be a large number of branches. Upon the screen the appearance will be as if a tree were growing; if at short distances apart in the tank drops of different colors are placed, the branches will interlace and produce a fine effect. A tank of coal-oil, in which is dropped a little colored fusil oil, is said to produce an entirely different figure.

But it is with the *vertical attachment* that the most novel and interesting phenomena, due to cohesion, may be shown. For this purpose it is necessary to have a horizontal tank, made by cementing a ring, an inch broad and four or five inches in diameter, upon a plate of clear glass. The ring may be made of glass, or wood, or zinc. This is to be placed upon the horizontal condenser, and half filled with pure water, the

surface of which is to be projected. Let fall, from a
height of two or three inches, a single drop of ether.
It assumes a characteristic form, will move about, but
will last only a few seconds, as it evaporates rapidly.
Rinse out the tank, and fill again with pure water, and
in like manner drop upon its surface any of the essential oils, of creosote, lavender, turpentine, sperm, and
colza oils. Each one will assume its peculiar form due
to cohesion.

Fig. 29 represents the pattern exhibited by a single
drop of oil of coriander, and Fig. 30 the appearance

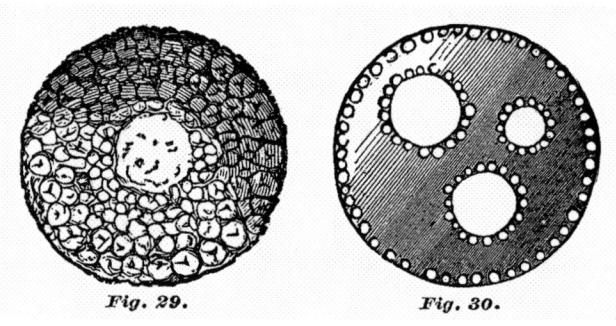

Fig. 29.   Fig. 30.

of oil of cinnamon. Some of these forms are very
beautiful, as, for instance, that due to oil of lavender.
This method of studying oils is used, by some experts,
to determine their kind and purity. These forms are
known as Tomlinson's Cohesion Figures.

Again, into the same tank, well cleaned and filled with
water, drop a few small pieces of camphor-gum. As
soon as they touch the water they will begin to move
rapidly, dodging each other in a wonderful way, and appearing as if they were endowed with life. Their movements will be accelerated if the water is warmed to a
hundred degrees, or a little more.

A drop of a solution of camphor in sulphuric acid, gently delivered to the surface of the water, will take a double-convex lens shape, and will move about the water in an eccentric manner for a long time. Several drops may be placed upon the water at a time, but they will avoid each other in their movements. Make a small boat of tin-foil, and into it put a fragment of camphor about the size of a pea, and place it on the tank; it may move round slowly, but put a piece of camphor, about the size of a canary-seed, upon the water, and it will spin round, dart up to the boat, and drag it about in a lively manner, just as an insect might do.

To show the existence of the camphor-film, that forms upon the surface as soon as it touches it, dust the surface of the water with lycopodium, then gently lower a fragment of camphor upon the middle of the tank. The instant the camphor touches the water the dust will be seen to open out into a circle of large diameter; then, after a moment's pause, the lycopodium is formed into a number of wheels, arranged in pairs, revolving in opposite directions.

A large drop of camphor dissolved in benzole, dropped upon water, has the appearance of a double convex lens; it sails slowly about for a while, becoming flatter and thinner, till at last it has sudden contractions, assuming different shapes. The contractions multiply till at length they become so violent as to throw off portions of the disk, or split up into smaller disks, which, in their turn, twist and double up, and ultimately throw out from each a tiny film of camphor, which lies quiet upon the water.

One who is interested to pursue this subject further will find an abundance of material by Tomlinson, in the *Philosophical Magazine* for 1861. Also in " Experimental Essays," Weale's Series, No. 143.

CAPILLARITY.

The chemical tank (Fig. 20), containing a little colored water, may be projected in any convenient way. If a small glass tube be placed vertically in the tank, the solution will rise in it. A series of five or six tubes, with bores of different size, may be placed in this tank at the same time, and the whole projected. The water will be seen to rise higher as the tube is smaller. A plate of glass three or four inches square may be put down into the tank, bringing one of its edges against one side of the tank. The water will rise two or three inches where the glasses touch, and slope away with a beautiful curve, which will vary as the whole side of the glass is nearer, or more distant from the other one.

CRYSTALLIZATION.

It is always fascinating to watch the growth of crystalline forms, especially when the process can be leisurely studied over a surface fifteen or twenty feet square. In all cases, a high magnifying power will be needed. Three hundred or four hundred diameters is better than any less.

If this is to be shown by a lantern, it will be necessary to have a powerful light, and the attachment known as the microscope attachment (Fig. 31), which fits upon the lantern (Fig. 26) when adjusted for horizontal projection. The lens must run forward nine or ten inches, and the jet drawn back until

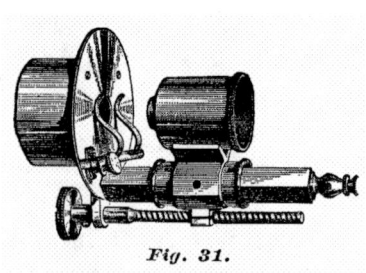

Fig. 31.

the maximum of light goes through the objective, which

has a short focus, and will not be more than three-fourths of an inch in diameter. A strip of clear glass an inch wide and three or four inches long, will answer upon which to spread the solutions to be examined, a few of which are given in another place. The glass will then only need to be placed in its receptacle, and its front focused, the same as for any microscopic objects. Any further instructions that may be needed, may be found under the descriptions of the method with the solar microscope. With the *porte lumiere,* and two lenses of proper focal length, the finest effects can be shown.

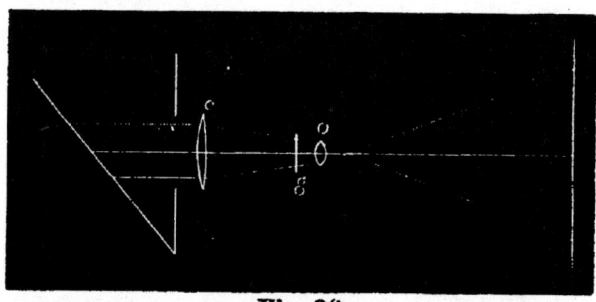

Fig. 32.

Let *c* be the condenser, with say twelve-inch focus, *o* the objective with one-inch focus; it may be a common pocket lens, or a linen prover, or a botanical glass. First adjust *c* so as to give a disk of light upon the screen. The rays will cross at the focus, and diverge afterward. Place the lens *o* so that all the light may pass through it, or as much as possible; this will depend upon the size of *o*. At any rate, it will be near the focus of *c*. Have ready a slip of glass three or four inches long and an inch wide, and wet one side with the solution to be crystallized; as, for instance, ammonium chloride, sometimes called sal ammoniac. Place it back of the objective at *g*, and move it until the wetted surface appears very plain upon the screen. Then wait until the solution begins to evaporate, as it will, from the upper edge first, when crystallization will begin there. See to it that the focus is right, and then gently blow upon the plate, unless the work is going on fast enough. The crystals will shoot out and grow while one looks, until they cover the entire screen with beautiful forms.

The following are good substances for illustration when dissolved in water: Ammonium chloride; barium chloride; copper sulphate; camphor dissolved in water; common alum; urea dissolved in alcohol.

ICE FLOWERS.

To exhibit the decrystallization of ice, which was first shown by Tyndall, it will be necessary to saw from a very clear piece of ice a cake three or four inches square, and about a half or three-quarters of an inch thick, cut parallel to the plane of freezing. When first cut, the sawn surface will be too rough for use, but will quickly melt smooth enough by dipping a few seconds in water. The beam of light that falls upon it should

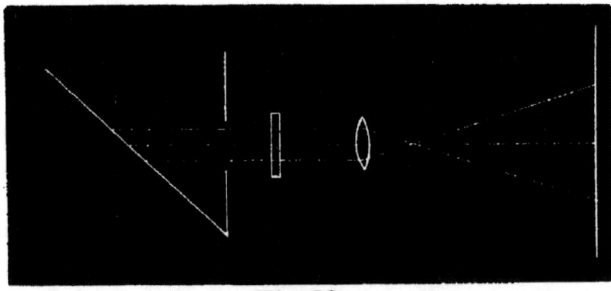

Fig. 33.

consist of parallel rays, and the *porte lumiere* is better for projecting this experiment than any lantern.

A single lens for an objective, four or five inches

focus or longer, will answer. It is the *interior* of the ice that is to be projected, and as there is a multitude of planes within it, each one being slowly decomposed, the light will suffer refraction, and one must not look for such plain figures to cover the screen as is represented in Tyndall's work, and in Deschanel's Physics. The forms can be picked out here and there.

If a lantern be used to project these crystalline forms, remember that the best effect will be obtained with a beam of parallel rays, which, in most lanterns, will necessitate the removal of the front lens of the condenser.

THE LEAD TREE.

Fill the small glass tank for the solar microscope with a rather dilute solution of the acetate of lead ; adjust it as for the exhibition of animalcules, using a small lens with a short focus, not more than an inch, if such an one is possessed. Into the solution now drop a very narrow strip of sheet zinc, not bigger than a common sewing-needle ; such a piece can be easily enough cut from a sheet of zinc with a pair of shears. This will at once have a deposit of lead upon it in a beautiful fern structure, which, while you look upon it, grows to be a forest. The same effect can be produced in the larger tank, described farther on, with a smaller magnifying power, by using a small battery of two Grove's cells, and having fine platinum wires to dip into the solution of lead. The lead will be deposited in the fern-form upon one of the wires. After there is a growth of the crystals upon a wire, attach the other end of the wire to the other pole of the battery, and then, completing the circuit again, the lead will be dissolved from the first, and be deposited upon the second.

THE TIN TREE.

Take a rather dilute solution of chloride of tin, made by dissolving the crystalline proto-chloride in water, in the proportion of one part of the former to four or five of the latter. This solution will precipitate its tin upon a piece of zinc in the same manner as the lead solution will, but the form of the crystals is very different. Use the same tank, and a magnifying power of 400 or 500 diameters, if good sunlight can be had. The growth will be quite rapid, and crystals six or eight feet long ought to appear. This needs no battery. Solutions of any degree of concentration can be used, but the growth is so rapid in very strong solutions, that the masses interfere with each other, and are dense and imperfect in form. Solutions can be used that are as dilute as twenty or more parts of water to one of the crystalline chloride.

THE SILVER TREE.

A solution of nitrate of silver is put into the tank, and a piece of fine copper wire put into it, the wire being nicely focused upon the screen. Pure silver will be immediately deposited in arborescent forms upon the wire, but the forms will vary with the strength of the solution. The more diluted it is, the finer will be the threads of silver.

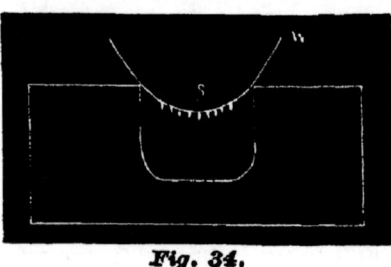

Fig. 34.

It is better to place the metal w that is to have the deposit upon it, whether of copper or zinc, so that it is just below the surface (s) of the solution, for the reason that when it is projected it is inverted, and as the arborescent deposit *hangs* upon the wire, it will appear upright upon the screen, and so have a closer semblance to a rapid vegetable growth.

A neutral solution of the terchloride of gold will give a characteristic growth upon a piece of zinc, but the solution should be quite weak.

Salts of copper will give nodular forms upon zinc, if very dilute, and a dense fringe of black copper, if the solution be very strong, sometimes terminating in quite large crystals.

GRAVITATION.

Make a frame like the picture, consisting of two upright posts, about one foot long and one inch square,

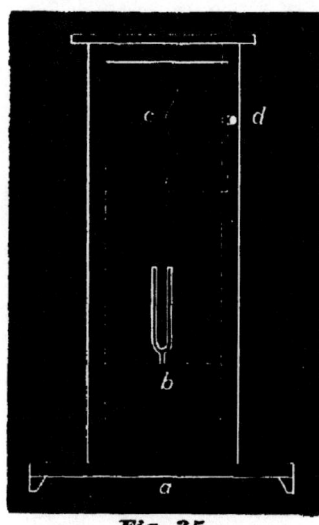

Fig. 35.

grooved like flooring on one side of each. Fix these into a board (*a*) about eight inches by twelve, for a support. Fasten a strip across the top to hold them steady. They should stand about five inches apart. The flange should be cut away from the right-hand standard, from the top down about five inches, so that the weight *b*, which has a tongue on each end, can be put into its place and be free to move up and down between the standards. A plate of glass (*c*) of proper width must fit in the top, and be fastened by a button (*d*), or otherwise, and held firmly in place.

Procure a pocket tuning-fork, either an A or a C, and

solder to one end a piece of small copper wire, so it will project from the side about one-eighth of an inch. A gimlet-screw can be cut into the other end of the fork, so that it can be tightly screwed into the wooden weight *b*, if small gimlet-holes are made before. There should be a number of these holes bored into *b*, at such a distance from the front edge that when the fork is in one of them, the wire upon the prong will come against the surface of the glass plate when the weight *b* is raised.

In order to use this instrument, it is necessary to coat the front side of the glass *c* with smoke, photographic varnish, or a *very thin* coat of white wax. Fix it in its place in the frame, and then raise the weight *b* (which ought to weigh two or three pounds) until the top of the tuning-fork is above the glass. Seize the two prongs of the fork with the thumb and forefinger, and pinching them close together, suddenly let it drop. The wire finger will trace a sinuous line upon the prepared surface, caused by the vibration of the fork during its descent. The undulations will be seen to increase in length as they approach the bottom; but as each one was made in the same time with every other one, it is obvious that the velocity increased as it was falling. In order to show this, it is merely necessary to put the apparatus near the condensing lens, and project the face of the glass. The line traced by the fork will be seen upon the screen. It will now be well to measure the lengths of the undulations, which can be very well done by having a scale in millimeters etched, or otherwise fixed upon another glass, which can be put just in front of the first, when the number of divisions of the scale to each undulation can be counted, and the result stated in mathematical terms.

PLATEAU'S EXPERIMENT.

With the vertical attachment and a tank, made five or six inches deep and with a plane glass bottom, this beautiful experiment, which so well illustrates cohesion and centrifugal force, may be projected. Fig. 36 shows

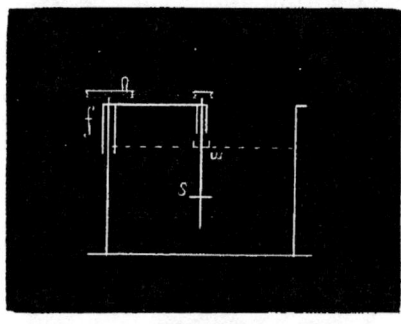

the proper conditions. A wire, w, is made to revolve vertically in the tank, by means of a little pulley driven by a cord about a larger one, at f, the whole so made as to rest upon the edge of the tank, and supported by ears,

Fig. 36.

as shown. The wire, w, should have a thin disk of tin fastened to it at s, for a surface of adhesion. Now the solution may be one of alcohol and water, so graduated that its specific gravity shall just equal that of the oil used, which can only be done by trial in a test-tube ; or it may be a solution of zinc sulphate, and the sphere may be made of bisulphide of carbon, with a little iodine dissolved in it, which will make it black, as in the former experiment described under the head "Cohesion." Here, also, the solution of zinc sulphate and water will need to be of the same density as the bisulphide of carbon, which will be best found by trial. This fixture must be placed upon the apparatus for vertical projection, and the focus adjusted for the sphere. If the above fixture for producing the rotation be made of stiff wire, it will not interfere much with the distinctness of the projection. A full account of this experiment, and of all the conditions to be observed, will be found in the *Smithsonian Report* for 1865, p. 207.

ACOUSTICS.

THE TUNING-FORK.

The vibrations of an ordinary tuning-fork may be exhibited in the following way. Having made the fork to vibrate, hold it at *a* in the divergent beam (Fig. 21), and swing it in its plane of vibration at right angles to the beam of light. Its shadow will present a curious, fan-like appearance. If the fork is polished it will reflect enough light to exhibit the same appearance when looked at while vibrating and swinging.

Another way is to hang light pith or cork balls so they just touch the fork, or other sounding body, and project the ball in any convenient way. As soon as the body begins to vibrate it will drive the ball away from it. Two forks in unison may be used in this way, to show sympathetic vibration. Hang a cork ball half an inch in diameter so it will just touch the side of one of the forks near the end, and project the ball and fork. At some distance set the other fork to vibrating, and put it upon its resonant case, or place the stem upon the floor or some resonant surface. The ball will be at once thrown off from the first fork, showing that it has been set vibrating.

Professor Mayer has described a number of interesting experiments to illustrate the change of wave-lengths by the motion of translation of the sounding body, in the *American Journal of Science, April,* 1872.

THE KALEIDOPHONE.

To the end of a piece of steel wire, two or three feet long, and an eighth of an inch in diameter, *o l* (Fig. 37), fasten with marine glue, or sealing-wax, a small bit of

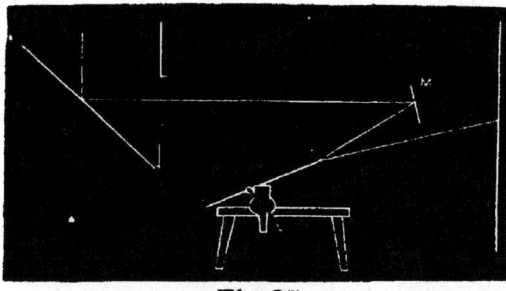

mirror, about the fourth of an inch square. The wire must be held tightly at some point, in a vice upon a table. The light from the

Fig. 37.

porte lumiere falls upon the plane mirror *m*, and is thence reflected upon the small mirror on the end of the wire at *l*, whence it is reflected to the screen. If the wire be now carefully plucked, it will give a line of light upon the screen, but will probably soon change into an ellipse or a circle. If the wire be struck with a small billet of wood, like a hammer-handle, there will be heard two sounds, the fundamental with some over-tone that will give a beautiful compound figure upon the screen, some circle or ellipse made up of small undu-lations, which will vary as the wire is struck in different places. If the wire be made fast at its middle, and the other end of it be plucked, the end with the glass will take up the vibrations at once — a case of sympathetic vibration. If it is not fastened in the middle there will be little or no movement when the lower end is struck. (See Tyndall on *Sound,* pp. 133, 135.)

MELDE'S EXPERIMENT.

To one prong of a small pocket tuning-fork tie a piece of silk thread, six or eight inches long, and to the other end tie a pin-hook and hang upon it a small weight, say a shirt-button. Project this with the large

lens, as represented in Fig. 38. First, with the fork
held as indicated, make it to vibrate. The string will
divide up into segments, all of which can be plainly seen
and counted. Second, turn the fork so that it vibrates

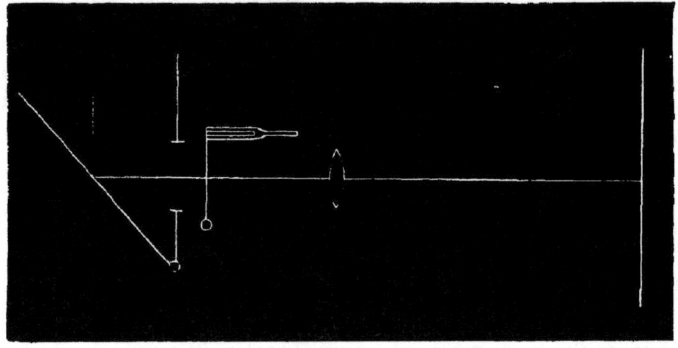

Fig. 38.

in a horizontal plain. The number of segments will
be doubled. Third, hang another button upon the pin-
hook, so that the weight will be doubled. Count the
segments while the fork vibrates, both perpendicularly
and horizontally. In this way some of the laws of
vibrating strings can be demonstrated.

Fasten a small piece of wire to one prong of the tuning-fork, and when the latter is vibrating draw it quickly
across a piece of smoked glass. The undulating line
will show well when projected.

THE OPEIDOSCOPE.

Take a tube, of any kind, that is five or six inches
long and an inch or more in diameter, tie a thin rubber membrane or a piece of tissue-paper over one end,
and on the middle of the membrane glue a piece of
looking-glass that is not more than the eighth of an
inch square. The light from the *porte lumière* falls

upon a mirror *a*, and is received upon the bit of mirror upon the end of the tube. The open end of this tube is to be held at the mouth and various sounds produced, varying in pitch and intensity. The vibrations of the membrane will move the mirror, and the beam of light

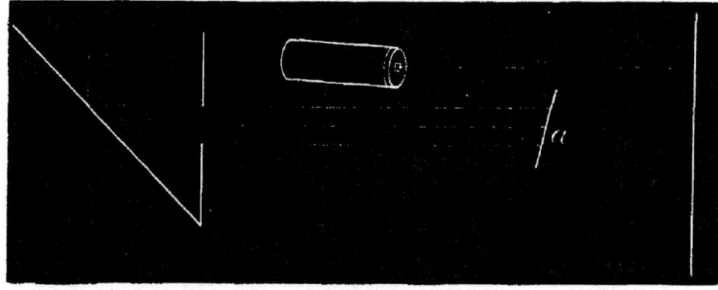

Fig. 39.

reflected from it upon the screen will describe various beautiful and regular curves, depending upon the management of the voice. It will be easy to find some pitch and intensity which will give a straight line : then, while the sound is being made, if the outer end be swung sidewise at right angles to the line, an undulating line will appear, in every way like those produced by the vibrating tuning-fork described on another page. If there are prominent over-tones in the sound they will be made apparent by their interference, giving a trace just like the traces upon a smoked glass by Scott's Phonautograph. The forms are regular enough for a tone of a given pitch and intensity, to enable one to write his music with them for notes; and if a tune like "Auld Lang Syne " be *tooted* in the instrument, the effect is quite amusing. The size of these figures, at the distance of fifteen or twenty feet, may be six or eight feet or more.

CHLADNI'S EXPERIMENT.

A glass plate of any form, if fixed by a clamp, will give out a musical sound when a violin bow is drawn across its edge. If the surface of the glass be strewn with sand, the latter will be arranged in some symmetrical form. The glass plate may be prepared as for the magnetic phantom, and the sand fixed after its acoustical arrangement, and afterwards projected as an ordinary transparency. It is generally best to exhibit this phenomenon during the process of arrangement, and this will require the fixtures for vertical projection. The glass to be sounded is to be made fast, and so placed that as much as is possible of it is over the condenser of the vertical attachment; then the sand sprinkled upon it, and the focus adjusted for the upper surface.

When the bow is drawn, the sand is seen to arrange itself according as the plate gives out one sound or another, which depends upon the part of the plate that is bowed, and where it is damped, also upon its form. It is well to have round, square, triangular, and hexagonal pieces, eight or ten inches in diameter.

To show water-waves upon a Chladni plate, Professor Morton has devised the following way: A plate of glass about a foot square is so held by its middle that one corner covers the condenser of the vertical lantern. To this corner is cemented a thin ring of soft rubber, of about five inches in diameter, and into this water is poured to the depth of one-tenth of an inch. Project the surface of the water and then draw the bow across the edge of the glass, as in the other cases, so as to produce a musical sound. The water within the rubber ring is thrown into a system of large waves, which form a shaded net-work of singular beauty. Drawing the bow so as to produce notes of different pitch, the waves will be large or small as the notes are low or high, and with a mixed note it is possible to get two or more systems superposed.

If a common tuning-fork be struck and then have one of its prongs put in contact with the surface of the water in this tank, a beautiful radiation of ripples may be seen, resembling somewhat the arrangement of iron filings about the poles of a magnet. The motion of water in a shallow bell-glass can be projected by letting the parallel beam from the vertical lantern go through it, doing away with the condenser, as the vessel itself would act as a lens if water were in it. The bow may be drawn across its edge when it will give out a musical sound, the water will be thrown into ripples, and a large objective might be used to project the whole surface. The bell-glass may be filled with ether or alcohol, and then sounded. Some of the liquid assumes the spheroidal form, and these are driven over the surface to the nodal lines. (See Tyndall on *Sound.*)

MANOMETRIC FLAMES.

The flame of a candle, or lamp, or gas-jet, if a luminous one, can be projected upon a screen by using a

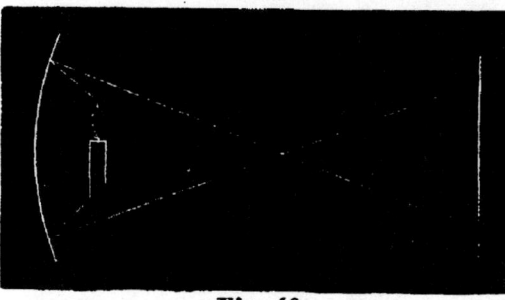

Fig. 40.

concave mirror (Fig. 40). It will be inverted and magnified. If while the flame is projected the mirror be tilted so as to swing

the beam horizontally, the flame will appear drawn out
into a band of light, due to persistence of vision. But
if the flame be not a bright one, the image will be too
dim to be useful, if the screen is ten or fifteen feet
away. The intermittent character of the singing hydrogen flame can be shown in this way, but it is much
better to use common gas in place of hydrogen, as the
flame is much brighter. The flame of common gas
may be made still brighter by passing it through benzole or naptha, or tow saturated with ether. The room
must be quite dark. (See Tyndall on *Sound*, p. 223.)
In the American edition of Atkinson's Ganot's Physics
is pictured Koenig's apparatus for observing manometric flames. In place of the rotating reflector use the
concave mirror, as above, and the same figures will appear upon the screen.

One can make a tolerable substitute for that apparatus,
if gas be not obtainable, by fastening over the mouth
of a small two-inch funnel, such as is used in chemical
laboratories, a piece of tissue-paper or thin rubber. A
piece of rubber tubing, two or three inches long, may
be drawn over the stem of the funnel, and the other
end drawn over the mouth of a common jeweler's blow-
pipe. A sheet of pasteboard may now be rolled so
large that the broad end of the funnel, which has the
tissue-paper pasted to it, may fit snugly in it. The
whole fixture may now be supported in any way, by
means of retort-stands. A gas-flame from a small
round orifice, or a common candle may be used for the
flame; the end of the blow-pipe is to be inserted in the
blaze, with the opening upward. If now, either a common mirror be used to give angular motion to the reflected beam, or the concave mirror to reflect the flame
upon the screen, while a sound is made in the large

tube, it will disturb the flame so much as to give a distinctly serrate image either upon the screen or in the plain mirror. The annexed figure will give an idea of

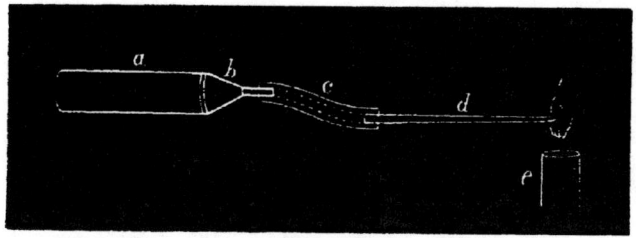

Fig. 41.

the arrangement mentioned: *a* is the tube for producing sounds, in *b* is the funnel with tissue-paper over its mouth, *c* rubber connection to the blow-pipe *d*, which opens upward into the flame from the candle *e*.

THE ORGAN-PIPE.

The vibrations of the *air reed* of a sounding organpipe may be shown, by having a small pipe made of iron gas-pipe and blown by illuminating gas, which may be lighted ; and when the pipe is sounding the reed will be seen to swing backward and forward in front of the *embonchure.* That it really vibrates may be seen by reflecting the light from a mirror upon a screen, and tilting the mirror, as is done in showing the manometric flames.

MACH'S EXPERIMENT.

The movement of the air within a sounding organpipe has been studied optically by Mach, a German physicist. His method was to stretch a membrane across the node of a pipe with glass sides, and in the open end he ran a fine platinum wire to the membrane, and thence out to be connected with a galvanic battery.

A sponge dipped in strong sulphuric acid was drawn along upon the stretched wire, the acid gathering itself up into small drops at regular distances apart. When

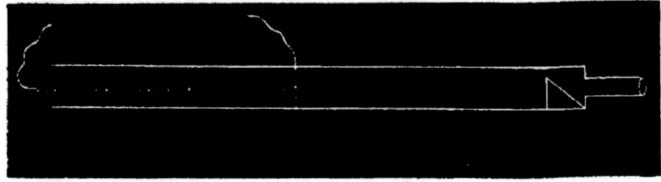

Fig. 42.

a current of electricity of sufficient strength was sent through the wire it was heated red-hot, and the acid was vaporized in dense fumes that, on account of its great density, sunk down toward the bottom of the tube, making so many gaseous strings hanging from the wire. These, of course, were subject to the motions of the air in the tube, and when the other end of the tube was sounded by wind from a bellows, the free end partook of the vibrations. The motions were then observed through a revolving stroboscopic disk, described further on. Not only the swaying of these gaseous threads was observed, but some of the Lissajous's curves were seen.

I think it highly probable that the motions of the air in such a sounding-tube can be shown to an audience, by having the tube with glass sides filled with dense smoke, and a strong beam of light converged in it, and having the stroboscopic disk so placed that the focus of the lens would be in the holes, and so permit a large amount of light to be used. Where the node was formed no movement would be visible; but by giving the disk a suitable velocity, at any other place than the node, the vibration might be shown in its different phases.

LISSAJOUS'S CURVES.

The optical method of studying vibrations is attractive to old and young, to students of science, and to musicians ; but the apparatus generally used is so costly that not many can afford to purchase it. The following directions will enable any one to have a pair of the tuning-forks made at the nearest blacksmith's shop, that will be found even more satisfactory for projections than the more costly ones.

Choose a piece of steel that is an inch broad, one-fourth of an inch thick, and about four feet and a half long. Have it made into two large tuning-forks, one of them to be about fifteen inches long, and the other twelve inches. Let the tines be two inches apart, and the flat sides should face each other on each fork. A stem may be now *welded* upon the bend ; it should be about five-eighths of an inch in diameter, three or four inches long, and made of round steel. When one of these forks is struck in the manner of common tuning-forks, it will be seen to vibrate through quite a large arc, and will continue to vibrate perceptibly to the eye, for half a minute or more. If, while the fork is vibrating the stem be held upon a table or floor, or some other resonator, a deep sound will be heard, and the larger one will make about fifty vibrations per second, while the shorter one will probably make seventy or seventy-five vibrations per second. A stand will be needed for each of these, and may be made by mortising a post three inches square, and three or four inches high, into a board eighteen inches long and ten inches wide (Fig. 43). This post should have an inch-and-a-half

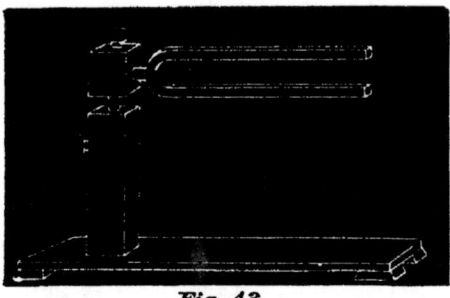

Fig. 43.

hole bored through it lengthwise, s o that a smooth stem may freely turn in it. T h i s s t e m must have a large head upon it, thro' which is bored a hole to receive the stem of the fork. Set-screws should be provided, to fasten the stems in their proper places. These supports might be made of cast-iron, in which case they would not need to be nearly so large.

Next make four slides of iron, an inch and a half or

Fig. 44.

two inches long, and bent so as to slide upon the fork and be fixed with a set-screw where it is wanted. These are for loading the forks and making them vibrate slower, as they are nearer the ends.

Lastly, each fork will need a small mirror fastened to its end. The small, round pocket mirrors, about an inch in diameter, I have found to answer well ; but care should be taken, in selecting these glasses, to get *plain* mirrors. Most of these small ones are on poor glass, and will spread a beam of light over a large space. These mirrors may be fastened to the end of the fork with the cement known as marine glue, and will adhere strongly enough for all careful work ; but sometimes these are fitted with a screw in the back, and screwed into a tapped hole in the end of the fork.

A still better way to fasten this small mirror, is to cement to its back a piece of rubber as long as the breadth of the fork, a quarter of an inch thick, and half an inch broad, this to be cemented to the end of the fork. The fork will not vibrate at all with this attachment at first; but if a thin wedge is cut out from each side of the rubber, until it moves very freely, the vibrations of the fork will not be much interfered with; at the same time the amplitude of the vibrations will be much increased.

When the mirror is fastened to each fork, it will be advisable to determine their pitch, which may be done by comparing them with a properly-tuned piano, organ, or another tuning-fork with known pitch.

EXPERIMENTS WITH THE FORKS.

I. The Sinuous Line. Cut off most of the light from the lantern or *porte lumiere* with a diaphragm, so that the beam is not more than an inch in diameter and consists of parallel rays. Adjust the fork so that it

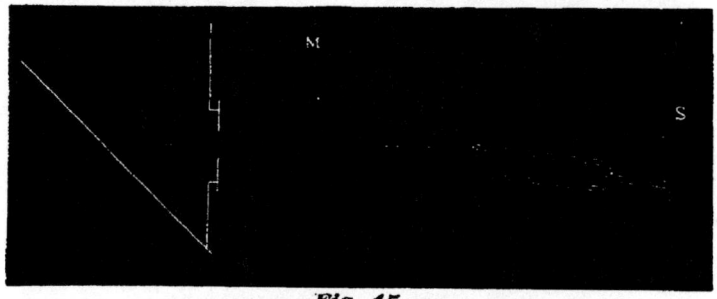

Fig. 45.

will vibrate perpendicularly, and place it so that the beam of light will fall upon the mirror at its end. This should be again reflected to the screen by a mirror *m* held in the hands, to swing the beam around the room. When the fork is made to vibrate by striking it with a small billet of wood, if the mirror *m* is held still,

a band of light will appear upon the screen, three to
five feet long, depending upon the amplitude of vibration and the distance to the screen. If now the mirror
m be turned so as to swing the beam at right-angles to
the band of light, a long sinuous line of light will be

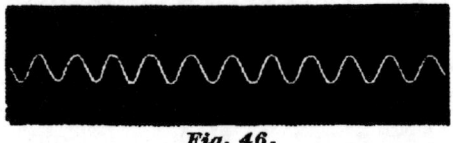

Fig. 46.

spread upon the wall. It may be seen to be forty or
fifty feet long if the mirror be moved fast enough. At
the time the fork is struck attention may be called to
the sound. If two beams of light, about half an inch
apart, and one above the other, be made to fall upon
the first mirror while it is vibrating, and the mirror *m*

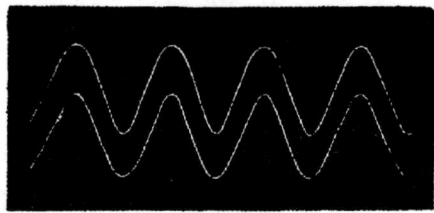

Fig. 47.

(Fig. 45) be moved as before, two undulating lines will
appear, one above the other (Fig. 47), with phases exactly corresponding. Let the two beams of light be

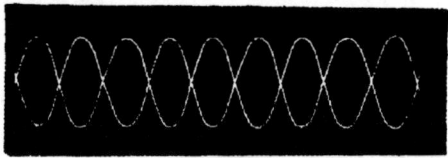

Fig. 48.

brought side-by-side and they will appear to have opposite phases (Fig. 48), and will show as beautiful interlacing lines. A double image prism put in the path

of the beam just in front of the fork, serves well to give this double line of light.

II. Overtone. If the fork be struck about midway of its length, a much higher sound will be heard along with the fundamental. Let the mirror be turning when the fork is struck, and the large sinuous line seen before will now be seen covered with ripples due to the overtone.

III. Interference. In the place of the mirror at *m*, place the second fork so that the beam of light from the first will fall upon it, and be reflected to the middle of the screen, having both forks to vibrate perpendicularly. Now load the shorter fork with slides until it is nearly in unison with the long fork. When they are both made to vibrate, the line of light upon the screen will be seen to lengthen and shorten with regularity ; at the same time beats will be heard corresponding with the lengthening of the line. These beats may be made to vary in frequency by moving the slides. If the beats are as many as five or six a second, or more, and the second fork be swung upon its vertical axis, the inter-

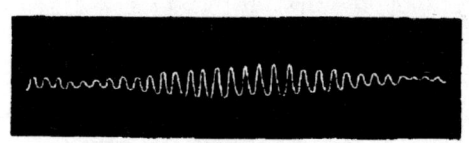

Fig. 49.

ference may be noted (Fig. 49) ; the swellings corresponding to the periods of coincidence, and the contraction to the periods of interference.

If the forks are now brought to unison and struck, the resultant figure will depend upon their relative phases. If they have like phases, so that each one reaches its limit at the same instant, the line of light upon the screen will be much elongated, the amplitude being equal to the sum of the two amplitudes. If their phases are opposite, so that one reaches its upper limit at the same instant that the other reaches its lower limit, then the spot of light will not be drawn out into a line at all, but will remain quiescent. These various relative vibrations can only be obtained by trial, but usually in four or five strokes one will develop such a phase as he requires.

IV. Resultants. Keeping the two forks in unison, turn the second fork so that it vibrates horizontally. Adjust the light so that it falls upon the second mirror as before, and thence to the middle of the screen. Now, if both forks be struck, the resulting figure may be a straight line, an ellipse, or a circle depending

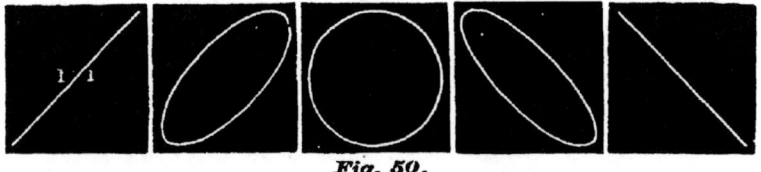

Fig. 50.

upon the phase of the first fork when the second one begins to vibrate. Fig. 50 represents these unison forms. By moving one of the slides so that the fork is not quite in tune with the other, the figure will move through each of these phases alternately. When the

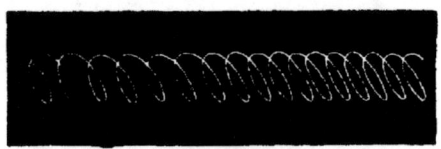

Fig. 51.

circle is obtained upon the screen, swing the second fork through a small arc, and the circle will be drawn out into a luminous scroll, (Fig. 51). If the forks are not quite in unison, the same experiment will give the scroll of irregular amplitude, (Fig. 52).

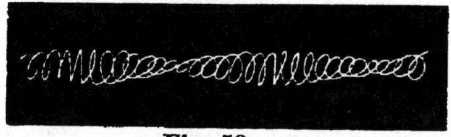

Fig. 52.

Remove the slides from the short fork and fix them upon the long one near the end, and, if necessary, attach two pairs, and adjust them so that the ratio of vibrations is as 2 to 1 ; that is, their pitch is an octave apart. The resulting figures are shown in Fig. 53 ; and

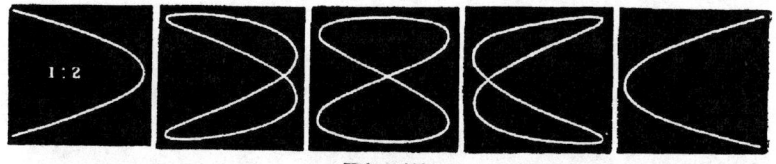

Fig. 53.

when the forks are tuned exactly, the figure first developed will remain, with no other alteration than a decrease in size, and may be a parabola, an 8, called a lemniscata, or an intermediate form.

While this figure 8 is upon the screen let the second

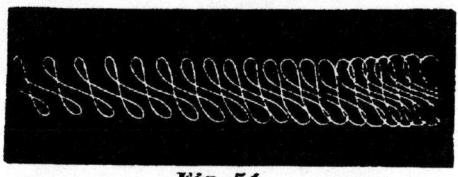

Fig. 54.

fork be rotated through a small arc, as before with the unison, and the scroll shown in Fig. 54 will appear.

By trial the slides may be so adjusted upon one of the forks that all the varying ratios in the octave may be obtained. The simpler the ratio the simpler the

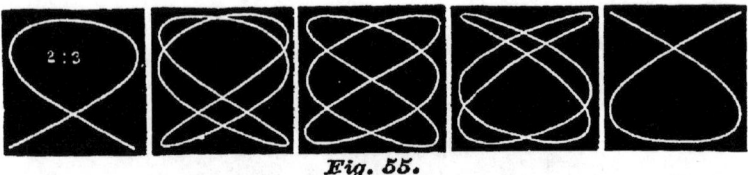

Fig. 55.

figure, and such ratios as 2 to 3 (*do* to *sol*), and 3 to 4 (*do* to *fa*), may be known by their representative figures,

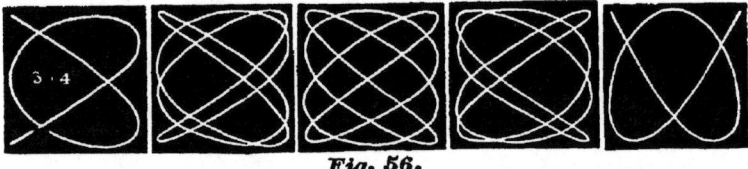

Fig. 56.

55 and 56. The ratio 1 to 3 (*do* to *sol*, in next octave,) will present such forms as those in Fig. 57.

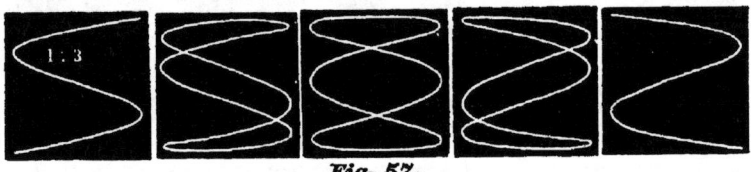

Fig. 57.

In any case, the figure will remain constant when the ratio is exact, and the ratio may be known by counting the number of loops upon the top and one side. Thus, in the *fully developed figure*, with the ratio 2 to 3, there may be counted two loops upon the top and three loops upon the side, which indicate that the fork that vibrates horizontally makes three vibrations, while the other one makes two.

The overtones may be developed and exhibited upon each of these compound forms by striking upon the fork rather lightly, about midway of its length, while it is giving any particular figure. Thus, if the forks are in unison and a circle has been obtained, the overtone developed will cover the circle with ripples which appear to move around it.

For the exhibition of the Lissajous curves with such forks as have been described, it is not necessary to use a lens, but the whole light from the *porte lumiere* may be allowed to enter the room, and the first fork placed with its mirror in the middle of the beam. If, however, it be desirable to admit less light into the room, a diaphragm may be used that admits a beam only an inch in diameter or less. A lens may be used which will concentrate the light upon a much smaller space, making a much brighter spot, but will very much reduce the size of the figures. When a lens is used, it must be so placed as to project the mirror upon the second fork. Its focal length should be two feet or more.

All of these phenomena can be shown by means of a lantern,—even an oil lantern will answer. It will be found best to use a beam of parallel rays, which may be used in such a lantern as is represented in Fig. 26 by simply removing the front lens of the condenser. With other lanterns it will be necessary to remove the objective, and push forward the light until the beam emerges with parallel rays: then, with a diaphragm cut off all the light except a beam of the size of the mirror upon the forks. The conditions are then the same as with sunlight, and a lens may or may not be used.

SYMPATHETIC VIBRATIONS.

Let the two forks be brought to unison and at right angles, so as to give, when struck, one of the forms of Fig. 50. If now, but one of the forks be struck, the straight line due to its vibration will slowly swell into an ellipse, which will be due to the absorption by the second fork of the vibrations of the first. This may be demonstrated by changing the pitch of one of he forks, when no change of form of the projected beam will be observed. One of the conditions for the success of this experiment is that both forks should rest upon the same table, in order that the vibrations may be conveyed through the solid wood from one fork to the other. The intensity of the sound-wave in the air is not sufficient to communicate a motion that will be perceptible. A voice sounding the same fundamental note as one of the forks, will set it vibrating, as will be evident by the spot of light upon the screen being drawn out into a line.

With one of these forks Melde's experiment may be shown in the most satisfactory manner. Choose a soft white cord eight or ten feet long (a silk cord is best, though a cotton twine will work very well), tie one end to the fork at *a* and let the other end hang over a hook driven in the wall at *b*. Weights varying from a pound to

B   **A**

Fig. 58.

half an ounce or less may be hung upon this free end of the string, with which its tension may be varied. The fork may be struck with a billet of wood, as in the former experiments, when the string will be made to vibrate, either as a whole, or in equal segments, the number of which will be inversely proportional to the stretching weight. The amplitude of these vibrations of the string will be considerable, and if the string vibrates as a whole it may be eight or ten inches, or even a foot; and when the number of segments is as many as sixteen or twenty, they can all be seen and counted by a large number of persons at a time. If the string *a, b,* is twice as long, and may reach back to *a,* the free end may be held in the left hand while the fork is struck with the right. It will then be very easy to vary the tension of the cord while it is vibrating, and the segments can be made to change through its whole series of one, two, three, four, etc. The various forms and motions of the cord may be shown to still better advantage, by making a strong beam of light from the *porte lumiere* or lantern to fall upon it in the direction of its length.

Crova's apparatus consists of disks of glass about fifteen inches in diameter, which may be made to turn upon a suitable rotator. These disks are at first painted black, and then curves of various forms are traced through the paint to the glass. The upper part of the disk is projected in the ordinary way, and then if it be rotated, the lines which are drawn upon it will appear to move or to be quiescent, according as they are concentric, eccentric, or some other form. If a diaphragm with a slit in it, long enough to reach across all the lines which are drawn upon the disk, be placed behind it, a series of dots will appear upon the screen, which will change their positions as the disk turns round.

With properly drawn curves the various wave-motions in air in organ-pipes, reflection of sound-waves, nodes, interference, and so forth, as well as the transverse vibrations in light-waves, may be well shown.

AN ATTACHMENT TO THE WHIRLING TABLE FOR PROJECTING LISSAJOU'S CURVES.

Two posts *p* and *p'* are made fast to the frame upon the opposite sides of the inertia plate *a.* A small wooden pulley *s*, about an inch in diameter, is made to turn upon an axis that is made fast in the post *p*, and with such adjustment that the pulley rests upon the

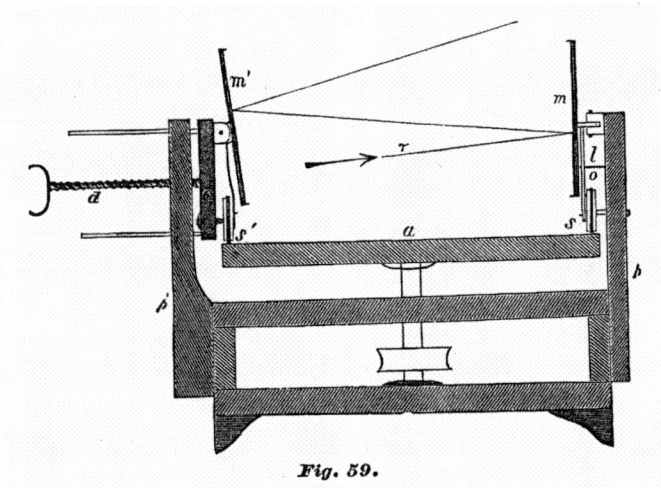

Fig. 59.

plate *a* and turns by friction on that plate. It is best to have a thin india rubber ring upon the friction pulley to insure it from slipping. Above the pulley the mirror *m* is so mounted as to swing in azimuth, and is made

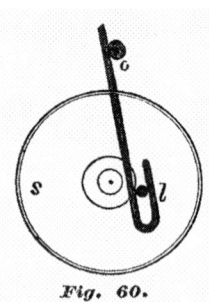

Fig. 60.

to do this by a wire fastened to it at its hinge, and bent into a loop *l* at its lower end, which is opposite the face of the pulley *s*. Another twist in the wire at *o* will be needed, for a pin which is fast in the post *p*; this will make a lever of the wire *l*, with the fulcrum at *o*, and if it is properly fastened to the hinge of the mirror will cause it to vibrate in a horizontal plane when the plate *a* revolves.

A somewhat similar arrangement is made for the other side, save that the friction pulley *s'* has its bearing made fast in a separate piece *c*, which is so fastened to the end of a long screw *d* that the whole fixture can be moved to or from the centre of the plate *a*. The piece *c* is furnished with two guides, which keep it steady in any place where it is put. The mirror *m'* is made to tilt in a perpendicular plane by an arrangement quite similar to the former one, save that the wire connection has its lower end bent into a horizontal loop, through which a pin in the face of the pulley *s'* is thrust. This is practically an eccentric, and, being directly fastened to the hinge of the mirror *m'*, gives to it an angular motion proportional to the distance of the pulley face-pin from the centre. The mirrors should be not less than two inches square. If then the pin is an eight of an inch from the centre of the friction pulleys, they will have ample angular motion ; much larger than can ever be got from forks.

Experiments.—It is evident that if the two friction pulleys have equal diameters, and they are at equal distances from the centre of the plate *a*, they will vibrate in unison in their respective planes. Now let a beam of light *r*, from the *porte lumiere*, fall upon the mirror *m* at such an angle as to be reflected first upon the mirror *m'*, thence to the screen. If the plate *a* is now revolved, the beam of light will describe a circle, an ellipse or a straight line, either of which can be made at will by simply adjusting the crank of one of the mirrors to the required angle. Thus, suppose the mirror *m'* is tipped back its farthest by bringing the pulley pin at the top, as indicated in the drawing, at the same time that the mirror *m* is at its maximum an gular deviation. The beam of light will describe a circle.

If it moves slowly, the path and direction of the moving beam can be nicely observed. These two advantages are not to be had with forks; for, first, it is accidental if one gets a circle or any other desired resultant figures from forks in unison, for the obvious reason that the phases cannot be regulated, and second, the vibrations of the fork are so rapid that the analysis of the motion can only be made in a mechanico-mathematical way.

By moving the fixtures on the left side toward the centre of the plate *a*, the pulley *s'* will not revolve so fast. If moved half-way it will make one revolution while the other makes two, and the vibrations stand in the ratio 1:2 represented by forks in octave. Such ratio is shown upon the screen by a form very much like the figure 8, and known as the lemniscate.

Between these two places, every musical ratio in the octave can be got, and the resultant motions projected in their proper curves. More than that, *while the mirrors are both vibrating*, any of the ratios desired can be moved to at once by merely turning the thumb screw *d*, which is wholly impossible with any forks which require stoppage and adjustment of lugs for each different curve.

Again, if the fixture *c* is moved still farther toward the centre than half-way, the curves projected will be those belonging to the second octave, until the pulley reaches three-fourths of the way, when the ratio will be 1:4, and the resultant figure will be like a much-flattened double eight.

If one would show the phenomenon of beats, it will be necessary to have the mirror *m* and its attachment so adjusted as to have it vibrate in a perpendicular plane like *m'*. This can be done by fixing its hinge at right angles, and the rest the same as for mirror *m'*.

The reflected beam from the second mirror may be received upon a large mirror held in the hands, and thence reflected upon the wall or screen.

LIGHT.

RECTALINEAR MOVEMENT.

That light moves in straight lines can be shown by admitting the light from the *porte lumiere* through a small hole. It goes straight across the room, and its course can be tracked through the room by the dust particles, or a little smoke, which it will light up. Also, by having the room otherwise quite dark, permit the light to come in the round orifice, half an inch in diameter, as it is reflected from the landscape outside, and not reflected from the mirror. The room is thus a large *camera obscura*, and an inverted image of the landscape will be seen upon the walls, or upon a small screen held a foot or two from the orifice. This image will be particularly strong if the ground be covered with snow, as much more light is reflected from that than from grass or foliage. If persons are passing their forms will be seen, and appear as if walking head downward.

Parallel rays A will be reflected from the mirror of

the *porte lumiere,* while *converging* B and *diverging* C rays will be obtained by interposing a convex lens of any size in the path of the parallel rays.

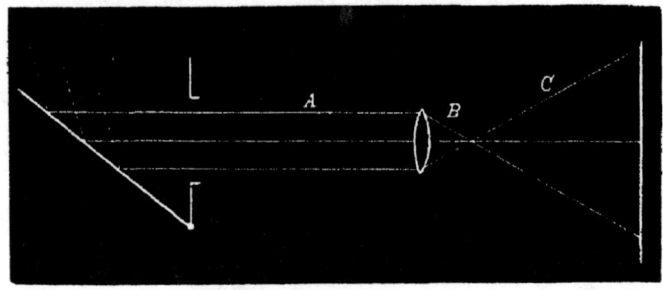

Fig. 61.

Transparent substances, like glass, some crystals, gases, and water permit the rays *a* to go through them and appear upon the screen. *Translucent* substances, like paper, ground glass, milk, allow but a few scattered rays to go through them, and a diffused light appears on the screen. *Opaque* substances, such as metals, thick pieces of wood, stones, etc., stop all the light, reflecting some and absorbing the rest.

INTENSITY OF ILLUMINATION.

When the lens is interposed in the path of the beam the light appears as a circular disk upon the screen, and as the rays cross each other at the focus *f,* that point may be considered as the source of light. Cut a sheet of paper or a board *s,* one foot square, and hold it any distance from the focus, say two feet. Its shadow upon the screen will be bounded by *a, c,* which may be measured in square feet. Now move the paper to *s′,* twice as far from the focus, and again measure the shadow *b, d,* it will be but one-fourth the size of the other, proving that at *s* the paper received

four times as much light as it did at *s'*. Hence the intensity of light varies inversely as the square of the
distance.   Other measures with other distances can be
made for confirmation : a good exercise for scholars.

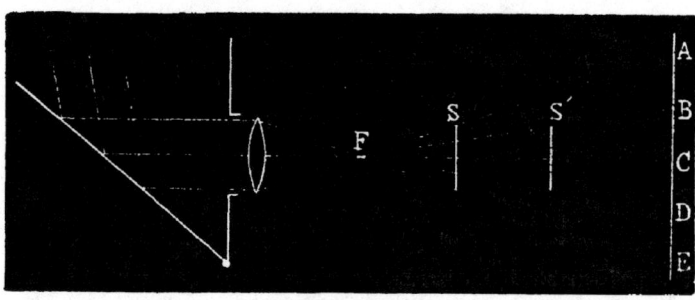

Fig. 62.

When a lantern must be used in place of sunlight, it
will be necessary to remove the objective and move the
light backward from the condenser until a sharp focus
is produced in front, and then work in front of that ; or
still better, remove both condenser and objective, the
outlines of shadows will be quite well defined with the
electric light, and with the lime light, but not with any
oil light.

REFLECTION.

The reflecting power of various surfaces can be
shown by holding them in the path of the beam from
the reflector. Common mirrors, plain glass, colored
glass, metals polished and unpolished, woods, horn,
polished stones, paper, will all exhibit difference in this
property.

Reflection from the two surfaces of glass is seen
upon the screen when the parallel rays from the first
mirror reach it. Then will always be seen two or
three indistinct images of the sun, side by side. When

the sun is near the horizon, so that the *porte lumiere* is nearly horizontal, more of these reflections will appear, due to multiple reflections upon the surfaces of the mirror. These can be magnified a good deal in the following way. Place the lens *o* at about its focal

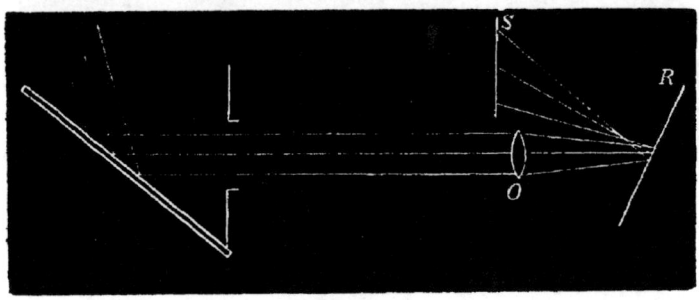

Fig. 61.

length distant from the orifice, and then hold another plane mirror *r* so that it will reflect the beam upon another screen *s*, moving the mirror *r* to such a place as to project the image of the orifice. It will be seen to be double, and when the images overlap, the light will be much brighter. Multiple reflections from the two surfaces of the mirror *r* may be seen by holding it at a small angle to the beam of parallel rays. A piece of plate glass two or three inches square answers for this experiment.

That the reflected beam moves through twice the angle of the incident beam, may be shown by holding the mirror *r* in the beam without the lens *o*. If the mirror be perpendicular to the beam, the light will be reflected back through the aperture; turning the mirror slowly when it is 45° to the incident light, the beam will be overhead 90°; when it has been turned 90°, and is now in the plane of the beam, the reflected part will have moved through 180°.

Pepper's Ghost is but a reflection from the surface of unsilvered glass. His fixtures were made upon a large scale, were costly, and not practicable in every place. His reflectors were large sheets of glass about five feet

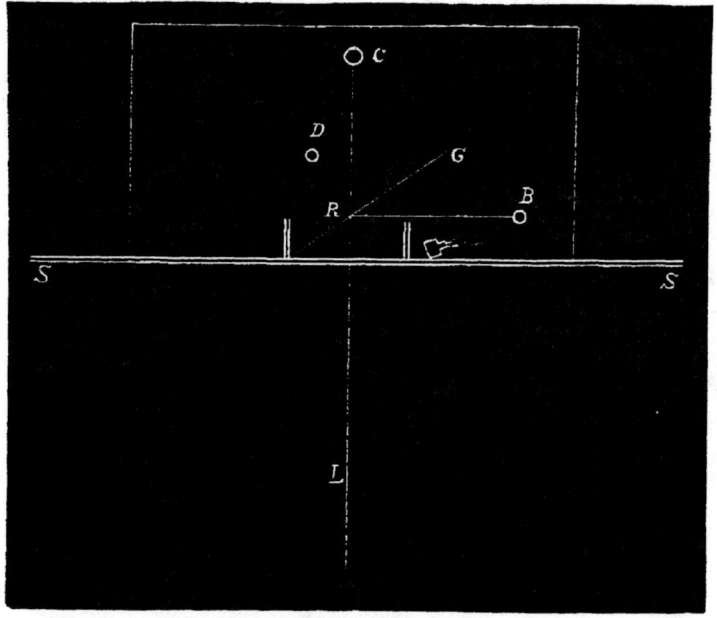

Fig. 62.

broad and six feet high. The light was a powerful lime light. Fig. 62 will give an idea of the conditions employed last year in his traveling lectures. The front of the stage *s s* was heavily curtained, except a space of a few feet in the middle of it, where there was a recess opening back, and apparently to the back of the stage *c*, which could be seen through a large plain glass reflector *g*, twelve or fifteen feet long and six feet high, placed at an angle of about 45°. This glass is seldom noticed unless one is looking for it. The lantern for illuminating the *ghost b* is behind the curtain on the right, and throws a powerful beam upon it. It being dressed in white, a good deal of the light is reflected from it in all directions, and a part of that which falls upon the glass at *r* will be again reflected toward *l*, when it will appear as if it came from *c*, as far back of *r* as *b* is front of it. All of the lights in the room are turned down except that in the lantern, and none of that is permitted to find its way into the room save what is reflected from the *ghost*. There is black cloth for absorbing the light back of *b*. The person who holds conversation with the phantom is at *d*, but of course he cannot see what those see who are at *l*, or near that line, and all his movements are guided by his knowledge of the necessary position of the reflection. In his book, *Cyclopaedic Science Simplified*, Professor Pepper has given several methods for showing such spectra. The skeleton, the talking head, and others are thus exhibited.

The extensiveness of the preparation for exhibiting the ghost will prevent most experimenters from attempting it; but if the teacher would care to show the principle, he will find the following a cheap and effective one, which he can extemporize with what materials

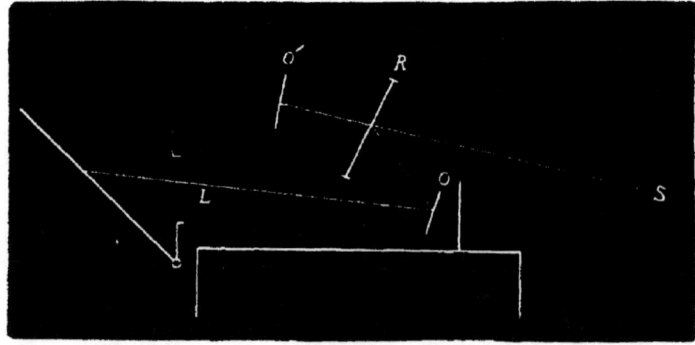

Fig. 63.

he is likely to have at hand. The beam of light from
the *porte lumiere* is directed upon the object *o*, which
should be a small one: a doll dressed in white, or even
the outline of one cut in white paper. The light from
it will of course be scattered from it in all directions.
A pane of white glass *r* will receive some of these rays,
and reflect them toward *s*, where they will appear to
come from *o'*. If the object *o* is a puppet or a moving
figure of any sort, it can be made quite a good phantom, though diminutive. The glass *r* can be moved so
as to give every one in the room a view of the phenomenon, while the hand put up to *o'* will reveal the shad-
owy nature of what is seen.

Of course all extraneous light should be shut out by
having the window curtains tightly drawn, and also
with black cloth about the apparatus to absorb all the
scattered rays, especially about *o* and *o'*.

Obviously, a lantern at *l* could take the place of the
sunbeam, but the light needs always to be a very strong
one, for but a fraction of the light is reflected from the
object, and this is again largely reduced by transmis-
sion through the glass; nevertheless, as the light is
used at the distance of but a foot or two from the object, it can be lighted sufficiently well for a small room
in the night with an oil lantern like Marcy's Sciopticon.

THE KALEIDOSCOPE.

The very great beauty and variety of the forms seen in the kaleidoscope makes them very desirable objects for projection. The following method will be found efficient:

1st. With *porte lumiere.*

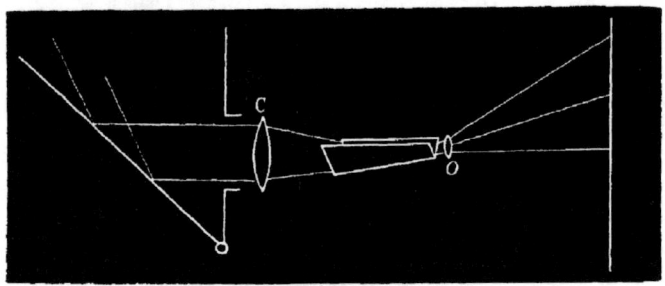

Fig. 66.

The condenser *c* may have a focus for parallel rays from a foot to eighteen inches or more. Choose an objective *o*, with focal length of eight or ten inches. It does not need to be more than an inch or two in diameter. Now cut two strips of looking-glass two or three inches broad and an inch shorter than the focal length of the objective. These may have the same breadth throughout, or they may taper to an inch broad at the outer end, as shown in the picture. They may now have their long edges brought together on one side and inclined to each other forty-five or sixty degrees, and secured there by enclosure in a tube ; or, if it be for temporary use, they may be held in place by a retort clamp and work just as well. The condenser *c* may now be placed close to the orifice and its focus will then be at some place *o*. Bring the fixed reflectors within the converging rays, so that they will receive the *focus just within their outer and narrower end,* and at the same time be so inclined that the light falls upon the surfaces of the glasses from the broader end, as shown above. Everything now depends upon the adjustment of the light to these reflectors.

When properly placed, there will appear, high upon the screen, the sectors of the polygon equally illuminated : six of them, if the reflectors are sixty degrees apart, and eight, if they are forty-five degrees. No *direct* light should fall upon the screen, and will not, if the end of the reflectors be kept high enough to cover the focus of the condenser. A few minutes' work with this will enable one to find the proper position for the best effect.

When the sectors appear equally illuminated, the objects to be projected may be placed between the condenser and the reflectors, — the fingers moved about, a pencil, a key, a comb, or a strip of paper with pins in it, or a leaf of a plant, perforated paper, or the common glass trinkets which are usually put in kaleidoscopes. If the objective lens be put close to the outer end of the reflectors, the objects shown will have a much sharper outline. For the best chromatic effects flat pieces of colored glass will be found better than round ones, as they transmit much more light, but an assortment of the two will make a fine appearance.

The common kaleidoscopes, which are so abundant in the market, can be used for this work by removing the ground glass in front of them and substituting a piece of plain glass. These are generally provided with a small lens, which will answer for an objective, but at the end for the eye there is seldom quite room enough to permit light to pass in sufficient quantity for good illumination. By removing the objective the diaphragm of black paper can be removed. As the objects are all magnified so much, it will be found that quite small bits of colored glass will look better than large ones.

2d. With a lantern.

It will be observed that the essential condition for showing the kaleidoscope with the *porte lumiere* is, that all the light that reaches the screen must be the light that is reflected from the inclined mirrors, and that the focus of the converging beam must fall just inside the outer end of them. Hence the focus needs to be as small as possible for the best effect. With the lime light the focus is quite broad at its narrowest part; therefore when the kaleidoscope is placed in the beam it will be necessary to adjust the light by raising it, so that the reflectors receive all of the light, and it also may be necessary to draw it back a little that the focus may come to the proper place.

The ordinary objective upon the lantern will not be needed, of course ; but an objective having a focal length equal to or a little longer than the length of the kaleidoscope may be used, holding it in a retort holder or in any other convenient way. Let as much as possible of the extraneous light be excluded from the room by black cloth about the front of the lantern. With these precautions a very good projection of kaleidoscopic forms can be shown. Even with the better forms of the oil lantern it is possible to project them quite well. As the diameter of the disk is doubled with this fixture, it will be necessary to move the lantern much nearer to the screen.

CONCAVE MIRRORS.

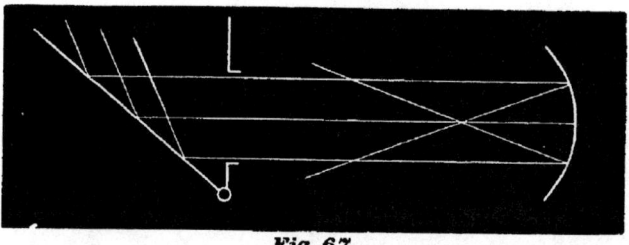

Fig. 67.

Concave mirrors, sufficiently good for demonstration, are fitted to wall lamps, or the reflector generally fitted to oil lanterns may be used. Such a one held in the path of the beam from the *porte lumiere* will reflect the rays to a focus, where there will be sufficient heat to ignite wood, paper, etc. If the mirror be tipped, the beam, after passing the focus, will diverge and cover the whole ceiling, as the focus is quite close to the mirror.

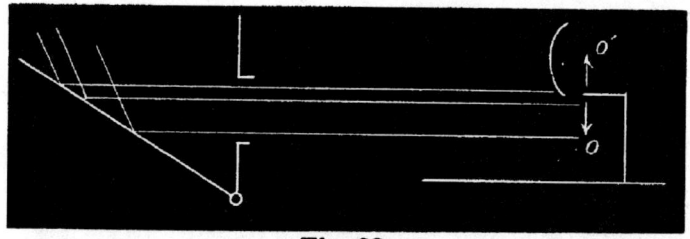

Fig. 68.

The image formed in front of the concave mirror may be seen by letting a strong light fall upon the object and having the mirror above it, as indicated. If the object *o* be inverted and hidden otherwise from view, it will appear upright; and at *o'*, to one standing in front of the mirror, all in a room can be made to see it by turning the mirror a little, so it will face them.

A small bunch of flowers, a statuette, the hand, etc., are good objects to exhibit this property.

Images can likewise be projected upon a screen by means of the concave mirror.

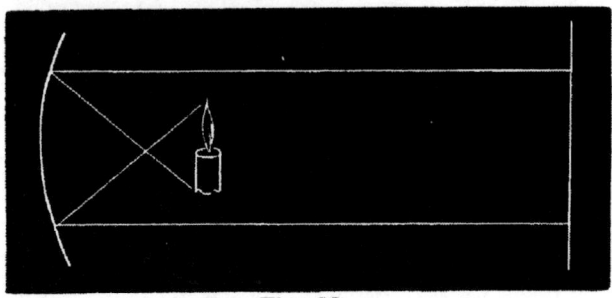

Fig. 69.

At a distance six, eight, or ten feet from a screen hold a lighted candle close in front of the mirror; slowly separate them: the image of the light will appear inverted upon the screen, and much enlarged. Advantage is taken of this property of the concave mirror to project some phenomena, such as manometric flames, etc., which see.

CAUSTICS BY REFLECTION.

A concave polished surface, like a strip of tin, two or three feet long and an inch or two wide, bent into

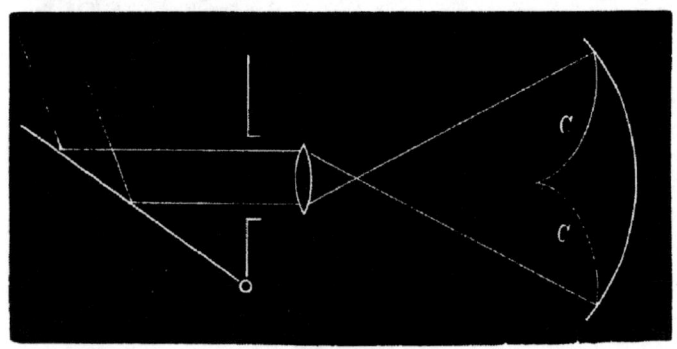

Fig. 70.

an arc of a circle or any other curve, held in the
divergent beam of light, as shown in the figure, and
resting one edge upon a white wall or a piece of
white paper, will exhibit fine caustic curves, which
will change as the strip is more or less bent. The
brighter the surface that reflects, the brighter will the
curves *c c* appear. Large rings, silvered and polished
on the inside, are sometimes used for this ; but a strip
of tin will answer well.

CONVEX MIRRORS.

The back of a concave mirror, such as already
mentioned, forms a very good convex mirror. Hold
that in the beam of light, in the same way as the concave mirror was held, and note the result. Objects of
any size are usually much distorted when seen by
reflection in a convex mirror, as witness your own
countenance when looking into one.

These distortions can be projected, though with
much loss of light, by strongly illuminating the object
o, and with an objective focus the reflection upon the

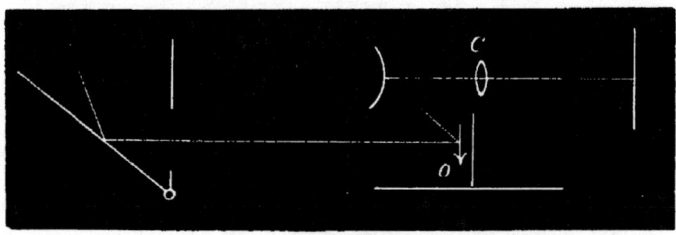

Fig. 71.

screen. In this way very humorous distortions of the
human countenance may be photographed by using
the camera at *c.*

TOTAL REFLECTION.

This phenomenon is generally shown by properly directing a beam of light into a vessel of water. Perhaps the simplest way is to fill a glass beaker with water, containing a little milk or a little magnesia stirred into it for the purpose of enabling the eye to trace the course of the light through it. Next placing the beaker in a convenient place, with a bit of looking-glass direct a small beam of light upwards through the side of the vessel, so as to strike the under surface of the water. By trial, the proper incident angle will be found at which the light will not emerge from the upper surface of the liquid, but will be totally reflected; the course of the beam will be easily traced through the milky fluid.

With suitable arrangements, very striking and beautiful effects may be produced in a stream of water.

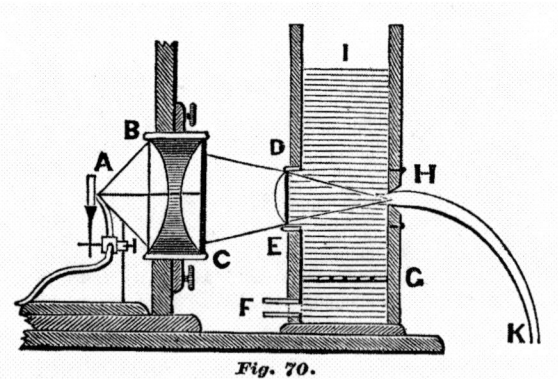

Fig. 70.

The high tanks made for showing the direction and form of water jets are generally made with a glass window opposite the orifice H, through which a beam of light from a lantern or from the sun may be directed while the water flows. For the success of this experiment it is necessary that the orifice should be round, smooth, and thin, and the body of water in the tank must not be disturbed by currents. In the figure, water is admitted at *F*, while at G there is a partition with a good many orifices in it through which the water flows, keeping it at a constant height I. When, therefore, the light is concentrated upon the orifice H, it is not scattered, but lights up the whole of the curved stream, giving it the appearance of molten silver. If colored glasses are interposed back of E, D, the color of the stream will also correspondingly change, with very pleasing effects.

Fig. 71 represents still another form of this experiment, in which the vertical attachment to the lantern is used. A vertical fountain jet is opened in the ascending beam from the lantern. The falling water is beautifully illuminated. Plates of colored glass may be used, as before.

MIRAGE.

Direct the beam from the *porte lumiere,* so that it is horizontal or nearly so. Put in a diaphragm with a hole about half an inch in diameter, or less, as the first condition is to have a small beam of parallel rays. No lens will be needed. Next heat a brick, or, still better, a poker or any convenient piece of metal that is a foot long or more, until it is nearly to a red heat ; then place it just in front of the diaphragm and parallel with the beam and about a quarter of an inch below it or to one side of it. The current of heated air will so deflect some of the light as to very much elongate the bright spot upon the screen, or even present another one some inches distant from the first.

Fig. 71.

REFRACTION.

1.—Of Glass.

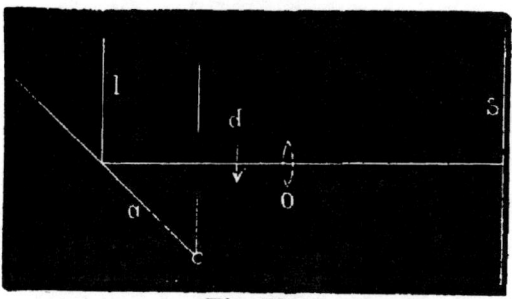

Fig. 72.

Project any object that is three or four inches long, — a lead pencil or an arrow cut out of paper. A single lens is all that is needed. Then hold in front of the object a piece of glass three or four inches long, half an inch broad, and the thicker the better. If the glass is held exactly perpendicular to the beam of light no refraction will be observed ; but turn one end of it towards the opening, and at once the picture upon the screen will appear as if a piece of the object had been cut out and was held to one side of it. The thicker the glass is the greater will be the displacement ; but a piece that is an eighth of an inch will quite likely make as much difference as the thickness of the object projected.

Two pieces of glass may be put together and held as before, or turned in various directions with reference to each other and the object.

2—Of Water.

A hand mirror held at *r* will reflect the light downward into the chemical tank (Fig. 73), which should be filled with water in which a little finely powdered resin

7

has been stirred to give a turbidity to it, as the beam will be traced much better ; a few drops of milk or of chalk-dust will answer the same purpose. A little smoke in the air will serve to mark the course of the

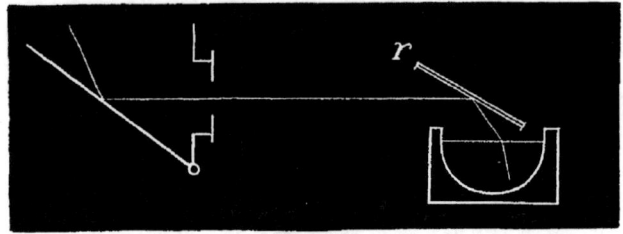

Fig. 73.

beam till it reaches the water, where its direction will be seen to change, becoming more perpendicular. Reflect the beam upon the water at various angles of incidence and mark the course of the refracted rays.

In place of the small tank take a beaker, or any vessel with square glass sides, and reflect the beam upon the surface of the water as before.

The experiment may be varied by filling the vessel half full of water, then carefully pouring strong alcohol upon it to the depth of an inch or two, being careful not to mix them while pouring, and some ether upon the alcohol in the same careful way. Sending now the beam through them as before, the different refractive powers of the various liquids will be seen.

3 — Heated Air.

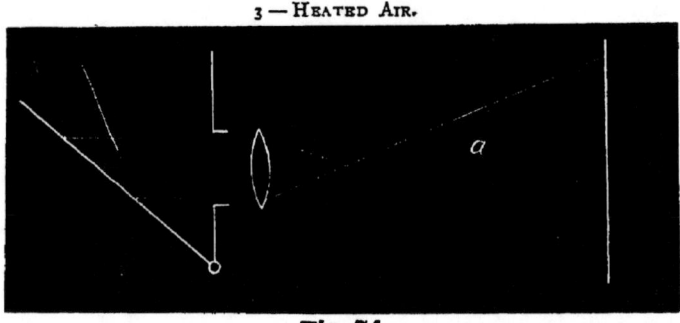

Fig. 74.

In the diverging beam from the lens hold a heated poker or any well heated body at *a.* The air that is heated in its neighborhood will so deflect the rays of light as to make quite a fine appearance of heat upon the screen. An alcohol lamp lighted and held there will have its whole projection upon the screen: the lamp and flame in outline shadow, while the heated gases rising make an interesting picture.

To present a still more striking case, project a large solar spectrum, as described on another page, and about half-way between the prism and the screen hold the lighted alcohol lamp, moving it slowly along in the different colors of the spectrum.

4—Lenses.

The refractive power of different forms of lenses may be shown by holding them in the beam of parallel rays. The refractive power of a lens of water is seen by taking two watch glasses or one watch glass, and a piece of plain glass a little larger, and bringing the two together under the surface of water. The space between these will be quite filled and they will adhere tightly enough for the experiment.

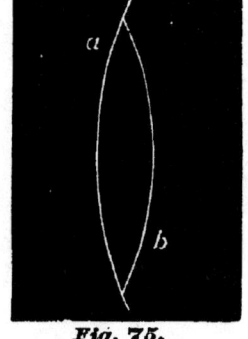

Fig. 75.

The same thing will also be exhibited by placing a watch glass upon the upper ring of the vertical attach-

ment (Fig. 27).   When water is poured in to fill it,
the light will be refracted and an object may be projected with it as with any other lens.

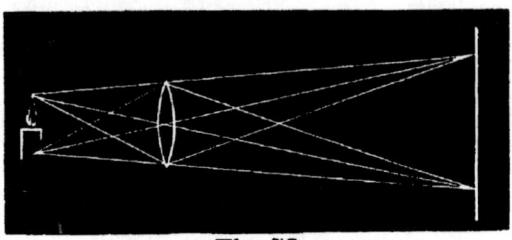

Fig. 76.

The formation of images by lenses is well shown by
holding a lighted candle or lamp in front of the screen
at any distance in a darkened room, and bringing the
lens close to the light, then moving it towards the
screen until the inverted image appears.   Try this
with double convex and with plano-convex lenses of
different focal lengths, also with a meniscus and with
a concave lens.

THE SOLAR MICROSCOPE.

This has been described on a former page, and may
be turned to ; but, as nearly all of the art of projection
depends upon the use of lenses, it will be well, in
giving instruction to dwell upon the conditions for
forming images with single and with compound lenses,
with parallel converging and diverging beams.   The
porte lumiere, the magic lantern, the solar microscope,
the telescope, may be illustrated by methods that have
been already explained.

THE RAINBOW.

This phenomenon in nature is due to refraction and
reflection in drops of water.   It is hardly practicable

to project a rainbow with an artificial shower, although it can be done by having the beam a widely diverging one, to fall upon spray from a small fountain that spreads a thin sheet at right angles to it; but a bow that will rival the natural one in the sky may be projected by using two lenses with short focus, such as are put into lanterns for condensers.

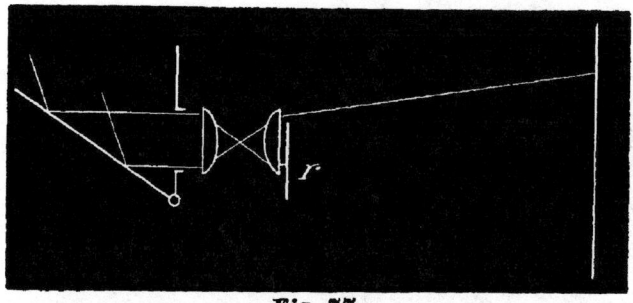

Fig. 77.

Place the first lens in position, as if for common projection. If the second lens be now brought near the focus of the former and slowly moved towards the screen, a luminous disk will appear upon it, having a red border. Let this disk be made as large as is desirable, which can be done by moving the lens backward or forward. Now cut a piece of paper *r*, with a round top a little smaller than the diameter of the lens, and place it at *r*. All the light from the lower part of the screen will be cut off, and nothing will be left but a bow, with the colors in the same order as those in the primary bow, and very brilliant. It may be enlarged to twenty or thirty feet.

A second method requires a conical prism; and, if this is not already possessed, one may be made by taking a thin, clear, glass funnel, with a mouth three or four inches in diameter. Cut a piece of plain white glass a little smaller than the mouth of the funnel, and fasten it there with wax or putty ; break off the stem ; put the prism under water, and when it is filled stop it with a cork : it is now ready for work.

Next cut out a semicircular slit from a piece of pasteboard. It need not be more than the one sixteenth of an inch broad, and it may be an inch across, as represented at A, Fig. 78.

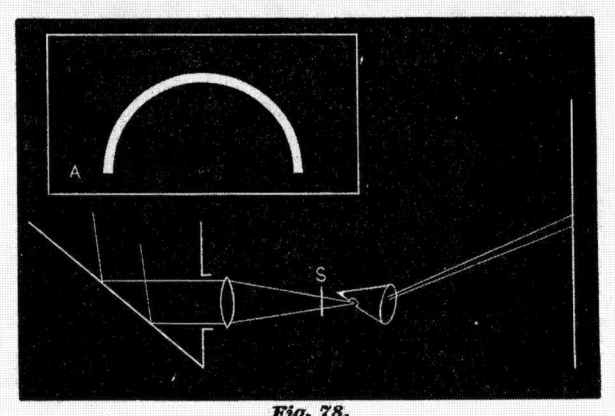

Fig. 78.

Place the condenser in front of the aperture, and hold the prism near the focus, and then a little distance back of it may be placed the pasteboard with the slit *s*. The semicircular diverging beam is refracted, and suffers dispersion ; which gives a very good bow, but with the red innermost, like the secondary bow.

Still another way is possible: If a beam of light fall upon a cylindrical reflector, like a glass rod, or even a tin tube, like the handle to a tin dipper, the light is reflected from it in a large nearly complete circle. Place such a reflector at *s* above, in place of the pasteboard slit, and then with the prism a bow will

Fig. 79.

appear, as before. The prism may be an ordinary one, but the bow will not then be perfect for a semicircle. The size of this bow will depend upon the size of the prism, for the ring of light can be indefinitely enlarged by varying the angle of incidence of the beam.

With a curved slit cut about the size of a common transparency and projected with the lantern, holding a common prism in front of the objective, a very good bow is seen. As the refraction bends the rays downward, it will be necessary to tip the front of the lantern up considerable in order to get the bow upon the screen. (Fig. 79.)

Lastly, let a beam of parallel rays, about an inch in diameter, fall upon a glass sphere filled with water, — an ordinary small glass flask answers well, but the larger the flask the greater should be the size of the beam. If, now, a small white screen be placed between the sphere and the aperture in the window, a bow will be seen concentric with the aperture and arranged so that the red is outside and the violet inside. At a greater distance from the aperture another bow will be formed, much fainter than the first, and with the colors in the inverse order. If different colored glasses are interposed in the path of the beam of white light, the bow will be seen to consist mainly of the tint of the glass.

CHROMATIC ABERRATION.

Caustic curves, due to chromatic aberration in the lenses, may be projected by taking two rather large lenses of short focus, such, for instance, as those made for lantern condensers.

Place the first one as if for common projections. The second may be held in the hand and brought near to the focus of the first and then inclined, as shown in the figure. Moving it towards the screen, beautiful colored figures will appear, which will change with the angle of the second lens to the light that is incident upon it. A comet, a hollow funnel, a mock sun, and

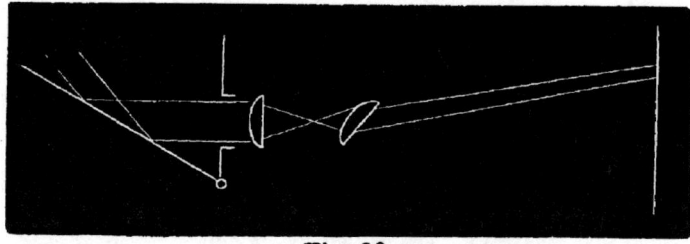

Fig. 80.

other curious forms may be projected, all of them brilliantly colored.

With a lantern, it will be sufficient to remove the objective and place a large lens, like one of the above, near the focus of the condenser, when the same figures will appear.

DISPERSION

Is usually shown by decomposing a beam of solar light with a triangular prism. The beam should be a rather small one, not more than one fourth of an inch in diameter, if a pure spectrum be wanted. If it be more than this, there will be more or less white light in the middle of the band.

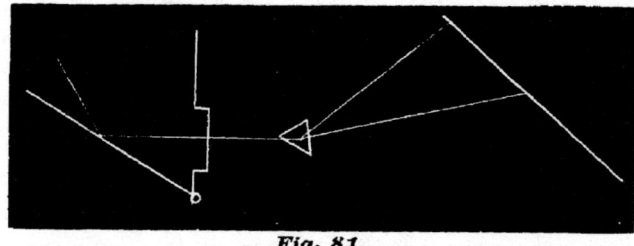

Fig. 81.

The smaller the aperture, the purer will be the colors into which the light has been decomposed; but if a very small beam be used, the room will need to be quite dark. When it is desirable to have a large and brilliant spectrum, the light may be sent through a condenser, with a focus one or two feet long, and using

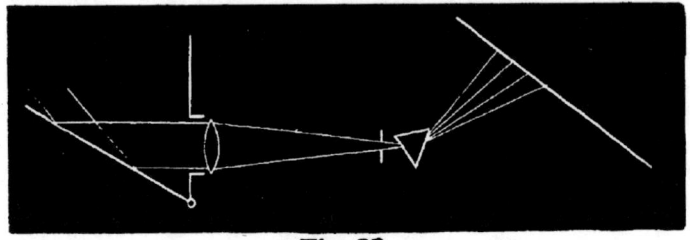

Fig. 82.

a diaphragm near the focus to cut off the marginal rays. This will permit much more light to be properly

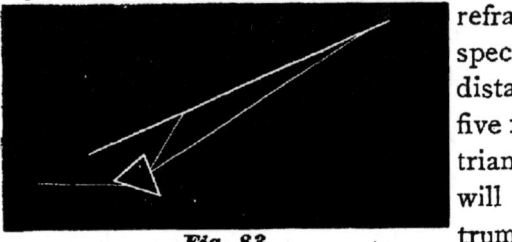

Fig. 83.

refracted for a good spectrum. At a distance of twenty-five feet, a common triangular prism will give a spectrum about five feet long; but it may be indefinitely lengthened by inclining the screen to it, as shown above, and it will usually be quite bright when made ten or fifteen feet long; if the room be otherwise, quite dark.

The dispersive power of different substances may be shown by making a V trough of glass, with an included angle of sixty degrees. Water, alcohol, ether, spirits of turpentine, etc., may be put in it, and the beam sent through them. In this case the spectrum will appear overhead upon the ceiling. If the glass trough have three or four tight partitions in it, all the substances may be used at once, and thus their refractive powers compared.

COLORS OF THIN FILMS.

Let a soap bubble be held in the beam of diverging rays, near the focus of the lens (Fig. 74), and in such a position that some of the light will be reflected from its upper surface.

As soon as the bubble becomes thin enough, brilliant colors will appear upon it, which will be reflected to the walls and ceiling, as they will spread over a large surface. If the bubble is held quiet long enough, each of the prismatic tints will appear in turn upon the walls, and sometimes the series will be repeated.

If the bubble is projected in the way mentioned upon page 44, three or four of these series may be seen at the same time.

Instead of blowing a bubble with a pipe, as shown in that figure, blow a mass of them in the dish containing the solution. Very large masses may be made and the colors reflected from them in the same way as above, or with the lantern.

The tension of the bubble film may be shown by leaving the tube open after the bubble is blown, when the latter will contract as if it were being drawn into the bowl of the pipe ; or the bubble may be blown upon the end of a glass tube bent twice at right angles, after which the open end may be put an inch or two under the surface of water in the chemical tank and projected.

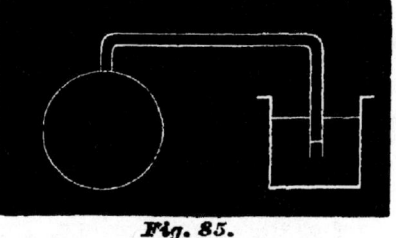

Fig. 85.

The water in the tube will stand below the level of the water in the tank indicating pressure.

When these colors from thin films appear upon the

screen, pieces of glass of various colors may be interposed between the lens and the bubbles, when dark or
black bands will be seen to take the place of those
colors that have been stopped by the tinted glass.

Yellow light that is nearly monochromatic can be
obtained by interposing a crystal of bichromate of
potash. Let the crystal be a thin and quite clear one.

Colored solutions may be used for the same purpose.

Under the head of Spectrum Analysis other means
for producing monochromatic light will be found, with
colored lights which are appropriate for examining
bubbles.

Bubbles made of common soap-suds will not last
long, and various preparations have been described
for making persistent bubbles, some of which would
last three days.

A piece of glycerine soap about the size of a marble,
sliced and dissolved in water at a 110° Fah., will make
a bubble that will last half an hour. Prof. Cooke gives
the following method for making a still more persistent
bubble : —

" Procure a quart bottle of clear glass, and some of
the best white castile soap (or, still better, pure palm-
oil soap). Cut the soap (about four ounces) into thin
shavings, and having put them into the bottle fill
this up with distilled or rain water, and shake it
well together. Repeat the shaking until you get a
saturated solution of soap. If, on standing, the solution settles perfectly clear, you are prepared for the
next step ; if not, pour off the liquid and add more
water to the same shavings, shaking as before. The
second trial will hardly fail to give you a clear solution.
Then add to two volumes of soap solution one volume
of pure concentrated glycerine.

NEWTON'S RINGS.

Choose a piece of white window-glass three or four inches square, and with clothes-pins or other means clamp it to the lens with longest focus you have; a lens with focal length of two or three feet will answer, though less curvature is better. Find by *rocking the lens upon the plate* with the thumbs where the point of contact is. This may be seen by a set of rings which surround it, and which move from place to place when the lens is rocked. Having found this place where the rings appear, place it near the focus of the condenser having a diaphragm of pasteboard with a hole in it not more than a quarter of an inch in diameter just back of the plate. This cuts off most of the light that would otherwise be scattered in the room, and prevents the rings from appearing plain. The objective used may have an inch focus. There will usually be seen as many as six rings, and the outer ones at the distance of twenty feet or more may be two or three feet in diameter. By interposing colored glasses or colored solutions, as with the bubbles, these colored rings will appear alternately with black rings.

RECOMPOSITION OF WHITE LIGHT.

This may be effected in several ways.

1st. By receiving the decomposed light from one prism upon the face of another prism like it, but turned so that the ray will have its original direction.

2d. By a lens. Let the decomposed rays from the prism fall upon a double convex lens placed so near to the prism that all of the colors of the spectrum may pass through it. Bring the screen to the conjugate focus of the lens, and then the light will appear as a brilliant white spot.   Interpose a piece of colored glass, and the spot will at once change its color.

3d. By reflection from a concave mirror. The colored rays will be converged as white light

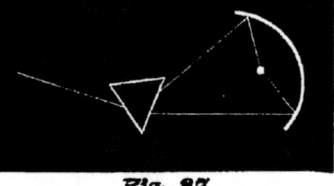

Fig. 86.

would be, and appear upon a small·screen placed at the focus as a spot of white light

4th. By reflection from a series of small mirrors. Let the spectrum fall upon the small mirrors, and so incline them that they will reflect

Fig. 87.

the light to the same place upon the screen or the wall.

5th.   By rotating colored disks.

Disks painted with the colors of the spectrum are sold in the market under the name of Newton's disks. They are made by pasting sections of colored tissue paper upon a large, stiff pasteboard disk.

These colors should have the following angular value : —

Red, 60°,	Blue, 55°,
Orange, 35°,	Indigo, 35°,
Yellow, 55°,	Violet, 60°.
Green, 60°,	

This disk may be rotated upon the whirling table, or, what is much better, a zoetrope rotator, and it will appear a dusky white.   It will be better to have a strong light thrown upon it while it is turning.

Another good way is to cut disks of properly-colored papers and make a radial slit in them.   When put upon the rotator, they can be made to slide by each other so as to expose a greater or less angle of any color. By using any two or more of these at a time, many interesting effects from combined colors can be exhibited.

One may often find colored stars or rings or other fanciful designs on posters for advertisements or wrappings on goods of various sorts, which may be utilized with the rotator in studying color.

FRAUNHOFER'S LINES.

The solar spectrum as usually projected with a round orifice and common prism, with an included angle of 60°, appears complete, and is often called a pure spectrum. If, however, the prism be of flint-glass or, better still, a bottle prism filled with bisulphide of carbon, it may be placed in such a position as to present the absorption lines known as Fraunhofer's.

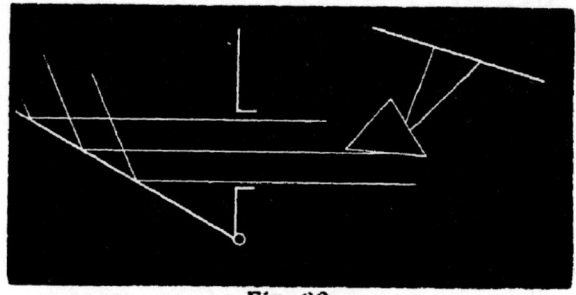

Fig. 88.

To do this it will only be necessary to place the prism in the full beam from the *porte lumière* and turn it so that one side is very nearly parallel with the beam. A spectrum will be formed containing a number of dark perpendicular lines known as the *C D E F* and *G* lines. These may be still more marked by placing a lens in front of the orifice at about its focal length distant from it, and placing the prism at its focus, and inclined to the concentrated beam in the same way as above. The spectrum will then be very bright and some lines well marked.

In order to show the Fraunhofer lines to advantage it is necessary to have the room quite dark ; to use a *very narrow slit* and a lens in conjunction with a good triangular prism of flint glass or of bisulphide of carbon. The diaphragm containing the slit through which the light must pass should be placed close to the opening in order to exclude all the light that is not directly used.

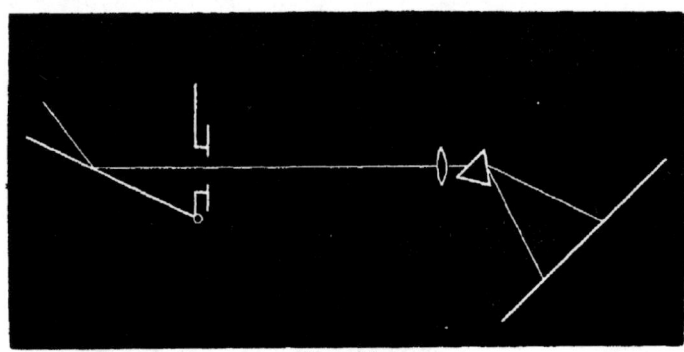

Fig. 89.

This diaphragm may be made of pasteboard with a slit cut in it three quarters of an inch long and the fiftieth of an inch in width ; the edges should be smooth and parallel. A lens with a focus of five or six feet is best for sharp definition of the lines, but one with a focus of only a foot or two may be used to exhibit the large and more prominent of them. Place the lens at such a distance from the slit as to project it sharply upon the screen, at a distance from the lens, say twenty feet. Then bring the triangular prism *close to the lens* as shown : The light will be deflected and dispersed, and the screen should now be brought where the spectrum will fall perpendicularly upon it, and at the same distance from the lens that it was before, namely, twenty feet. Turn the prism until the spectrum has its least deviation, which will be found by a little trial. The Fraunhofer lines should appear. If they are indistinct, move both the lens and prism back or forward in the beam until they are distinct, for it is now only a matter of focussing.

If the lens has a focus five or six feet distant it will need to be quite as far from the slit as the length of its focus, and the screen adjusted as before, but the lines should appear plainer and in greater number. With such a lens and a good glass prism the spectrum should be about five feet long, and with good focussing the D line should be seen double. These lines may be seen by a large number by moving the screen edgewise an inch or two.

One may use a condenser and converge a large beam upon the slit. This will make the spectrum brighter and permit a narrower slit to be used, but the definition of the lines is not so good as when parallel rays fall upon the lens. If the object be to project a spectrum that shall be well defined upon its sides and to show only the more prominent lines, let the slit be made as broad as the twentieth of an inch ; a lens with about a foot focus may be used to project the slit in the ordinary way, and the prism placed at the focus and turned to its angle of least deviation, which, as before, must be found by trial. In this way a beautiful and well-defined spectrum will be produced, which at the distance of twenty feet would be about five feet long and two feet broad.

8

ABSORPTION BANDS.

If a piece of colored glass be held in the path of the beam of white light before it enters the lens, Fig. 89, a part of the light will be absorbed and black bands of greater or less breadth will appear upon the screen. The glass may be held between the prism and the screen with about the same result. Some of the pieces of colored glass, which are quite common, will give very distinct absorption bands. It will be well to try red, yellow, green, blue, and violet glasses. If the color is very deep a greater width must be given to the slit else the spectrum will be seen with difficulty.

The chemical tank (see page 34) may be used to hold solutions of various kinds in this place. A wedge-shaped tank is also very convenient, as it enables one

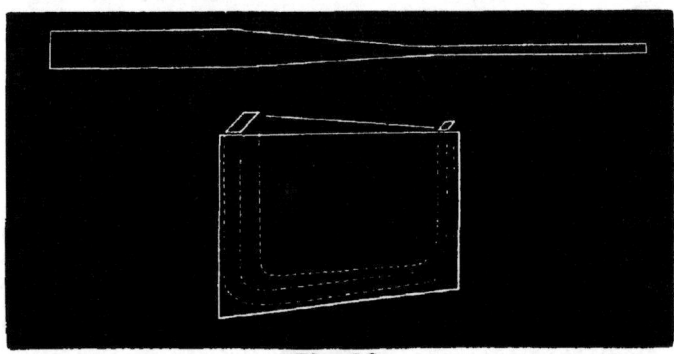

Fig. 90.

to pass the light through any required thickness of a solution, and thus to note the effects of thickness upon absorptive action. This tank may be made five inches long, four inches broad, and an inch thick at its broad end. A piece of thick rubber cut as in the figure will answer for bottom and edges of this tank.

Each end being bent up at right angles, the glass may be bound to it by clamps, as in the other tank.

" A solution of alizarin in carbonate of potassium or sodium, or in ammonia, exhibits a spectrum having a band of absorption in the yellow, another narrower one between the red and the orange, and a third very inconspicuous band coinciding with the line E. Purpurine dissolved in carbonate of potassium or sodium exhibits two dark bands of absorption about the green part of the spectrum. A solution of the same substance in aqueous alum exhibits the same peculiar mode of absorption, but likewise a yellow fluorescence. A solution of purpurine in sulphide of carbon exhibits four bands of absorption, of which the first, situated in the yellow just beyond D, reckoning from the red extremity, is narrower than the rest. The second is situated in the green, nearly coinciding with the line *E.* The third in the blue, near F, and the fourth, which is very inconspicuous, in the indigo. Lastly, the solution of purpurine in ether gives a spectrum giving two bands of absorption, one narrow and very dark in the green, nearly coinciding with *E.* The second in the blue, broader and less strongly marked, and having its centre at the line F ; the solution is also slightly fluorescent." (Stokes.)

The following series of experiments upon Absorption is taken from an article by A. H. Allen in Nature, vol. 4, p. 346. A lime light may be used if it is desirable to project these when sunlight is not available : —

" A beam of light from the lantern is passed through a slit, focussed by a lens, refracted by a bisulphide of carbon prism, and the spectrum exhibited in the usual way. A flat cell containing a solution of permanganate of potash is next placed in front of the slit. With

a weak solution and narrow slit a series of black bands
are produced in the green part of the spectrum; but
with a stronger solution the green and yellow are completely cut out, allowing only the red and deep blue
lights to pass. On widening the slit these bands of
colored light of course increase in width also, gradually
approaching each other until they overlap, producing a
fine purple by their admixture.

If the experiment be repeated, substituting for the
permanganate an alkaline mixture of litmus and potassium chromate in certain proportions, only the red and
green light are transmitted, the blue, and especially
the *yellow*, being completely absorbed.

On widening the slit as before, the red and green
bands overlap and produce by their union a very fine
compound yellow, while the constituent red and green
are still visible on each side. The effect is most striking when by the widening of the slit a round hole is
exposed in its place, when then appear on the screen
two circles, respectively green and red, producing
bright yellow by their mixture. This experiment is the
more striking as it immediately follows the process
of absorbing the simple yellow. The mixture above
described (suggested by Mr. Strull) answers better than
a solution of chromic chloride.

Of course, it is a well-known fact that all natural
yellows give a spectrum of red, yellow, and green, and
a common effect illustrating the compound nature of
yellow is noticed when exhibiting a continuous spectrum on a screen. When the slit is narrow the green
is very fully developed and only separated from the
red by a very narrow strip of yellow, while on gradually
increasing the width of the slit the red and green are
sure to overlap, producing the brilliant yellow we

generally notice. Thus the purer the spectrum the less yellow is observed.

If the continuous spectrum be produced with a quartz prism, a little management and adjustment of distance of the screen will cause the two spectra to overlap so that the red of one may be made to coincide with the green, blue, or any desired tint of the other. The same result is obtained by employing two slits at the same time, the distance between which can be adjusted. By this means two spectra are obtained simultaneously, any portions of which may be made to coincide.

A saturated solution of potassium chromate absorbs all rays more refrangible than the green, while a solution of ammonio, sulphate of copper stops all but the blue and green. These statements may be proved by placing flat cells containing the liquids in front of the slit of the lantern, and on placing one cell in front of the other in the same position, the green light only is transmitted. This experiment serves to explain the reason that the mixture of yellow and blue generally results in green, all other rays being absorbed by one or other of the constituents.

By placing the two cells in front of separate lanterns and throwing disks of light upon the screen, a beautifully pure white is produced when the blue and yellow overlap. I employ one lantern only for this experiment, using *two* focussing lenses side by side to produce the overlapping circles of light. I also employ a cell with three compartments, containing solutions of analine, ammonio, sulphate of copper, and a mixture of potassium chromate with the last solution, and projecting images on the screen by means of three lenses fitted on the same stand but capable of separate adjustment.

I can thus exhibit overlapping circles of brilliant red, blue, and green light, which produce a perfect white by their admixture ; while at the same time there is seen the compound yellow produced by the union of red and green, the purple arising from the red and blue, and a color varying from grass green to sky blue produced by the combination of the green and blue light. This experiment has the advantage of exhibiting at the *same time* the three primary colors, — red, *green*, and blue, — the compound colors produced by their mixture, their complimentary tints, and the synthesis of white light."

The flat cells mentioned are made by cutting thin pieces of board to the desired shape, and cementing pieces of window-glass on each side by means of pitch.

INTERFERENCE SPECTRA IN REFLECTED LIGHT.

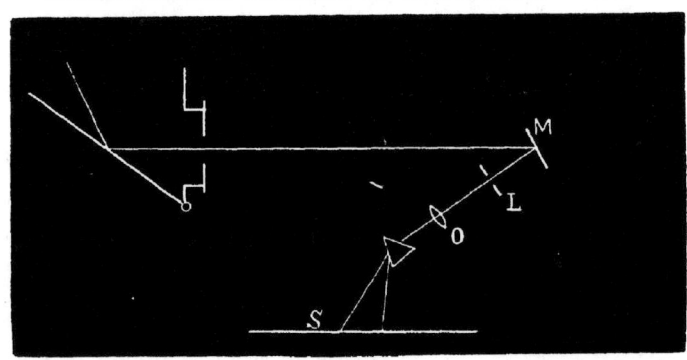

Fig. 91.

Let a beam of light about an inch in diameter fall upon a thin piece of mica, *M*, distant eight or ten feet from the *porte lumiere.* A part of the light will be reflected, and in that may be placed a slit at *L*, and a lens O may project the slit in the ordinary way. At the focus of the lens place a good prism so as to have a spectrum fall upon the screen at S. This spectrum will be seen to be traversed by a large number of black bands distributed throughout the whole length of it. If the plate of mica be very thin and white there may be as few as eight of these striæ, but if it be thicker their number will be largely increased.

The room will need to be made as dark as possible for this experiment, as the spectrum will not be very bright at best, and it therefore cannot be enlarged. If the length of the spectrum exceeds a foot it will be quite dim. These lines, however, can be seen to great advantage by placing the eye close to the prism when in its place as shown above.

If the spectrum of the light reflected from mica be received upon a paper screen painted over with a solution of quinine and thus rendered fluorescent, such interference striæ will make their appearance in the ultra violet part of the spectrum.

SPECTRUM ANALYSIS.

To project the spectrum of any substance whatever it must be heated until its vapor is brilliantly incandescent. The heat of the electric arc is best for this work as every substance is vaporized there. The lime light may be used to exhibit the principles of spectrum analysis, but its heat is insufficient for most of the metals. The characteristic lines of Sodium, Calcium, Lithium, Barium, Strontium, Potassium, and Copper may be tolerably well exhibited with a lantern furnished with oxyhydrogen jet and gases.

L1st. To exhibit the spectrum : —

Produce the lime light as you would for common projection. Remove the objective and place at the focus in front of the lantern the slit *d.* The objective *o* may then be so placed as to project a sharp image of the slit upon a screen in front of it at a distance of fifteen or twenty feet ; then place the triangular prism

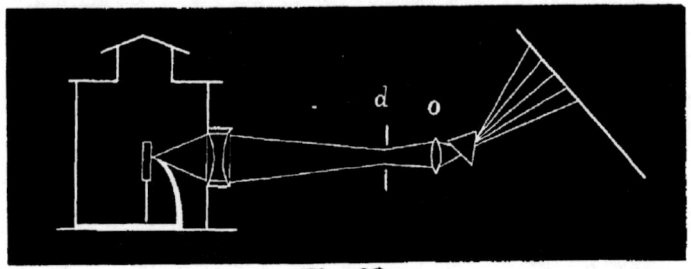

Fig. 92.

close to the objective. The screen will now need to be moved, that the refracted rays may fall upon it, and at the same distance from the objective that it stood in front, otherwise the edges of the spectrum will appear blurred. This should give a spectrum about five feet long at the distance of twenty feet, but the length will depend upon the dispersive power of the prism. It will be longer with a bisulphide of carbon prism than with one made of glass. If a still longer one is needed use two similar prisms close together and each one turned to the point of minimum deviation.

If a very pure spectrum is needed, all of the condensers may be removed and the slit put in their place. A parallel beam will then fall upon it, and the projection may then be made in precisely the same way as for the solar spectrum. In this case the light will be much less intense.

2d. To project the spectrum of the elements : —

Remove the lime cylinder and its holder, and light the gases : the tongue of flame will be six or eight inches long. Now hold a stick of glass like a large glass stirring rod in the flame at the same place when the lime cylinder is fixed: It will glow brilliantly with nearly the monochromatic light of sodium, and if the prism is in its place the bright yellow line indicative of that element will appear upon the screen. The glass will need to be turned slowly, and the attention of one person will be needed constantly to keep it in place. Sticks of soda glass may be had in the market, made especially for projecting the sodium line in this way, but the spectrum can be obtained from almost any piece of glass.

Another good method is to soak soft-pine sticks six or eight inches long and half an inch thick in saturated solutions of the *chlorides* of the various elements to be projected, as the chlorides are more volatile than other salts. Let the sticks remain in these solutions several days before they are to be used, as a much larger quantity of the material will be absorbed. These solutions may conveniently be made in test tubes six or eight inches long, remembering to label each tube by pasting a bit of paper upon it and writing the symbol of the substance contained in it. The chlorides of all the substances named above may be prepared in this way and a stick provided for each one.

The saturated and still wet stick must be put immediately into the flame where the glass and the lime cylinder are otherwise placed, and, holding one end in the hand, keep turning it slowly. The stick will glow and give out the kind of light that is peculiar to the included element.

The spectrum consisting of bright lines will be quite bright and sufficiently large to be plainly seen by an audience of a thousand persons. Sodium, Calcium,

Lithium, and Copper are especially good for this work and give satisfactory spectra.

When this monochromatic light from the stick of glass or the saturated solution of sodium chloride is made to appear, it will be a good time to give attention to its effects upon other colors. Observe the faces of individuals, the colors of flowers, of ribbons, of pictures. It is a good plan to have prepared a set of strips of bright-colored papers, or ribbons, or the Newton's disk, for exhibition in monochromatic light.

REVERSED LINE.

The dark sodium line is the only one that is ever projected, owing to the great difficulty there is in making the vapors of other substances sufficiently dense to absorb the powerful rays from the electric arc or of the lime light. With either, a pure spectrum must first be projected, and the slit should be nicely focussed, as described. — Then having provided a gas jet with Bunsen burner, or an alcohol lamp in front of the slit, hold in it a small iron spoon containing a lump of metallic sodium as large as a pea. It will take fire and burn with a yellow blaze and a white vapor, *through which the light from the lantern must pass.* If this vapor is dense enough it will stop rays from the other light that have the same refrangibility; and as its own luminousness is not very great, it will leave a black line upon the screen in the place where the sodium line would appear if the light came from it.

It will be best to have a screen a foot square with a hole through it, to set in front of the sodium flame to prevent its light from falling upon the large screen and injuring the effect.

FLUORESCENCE.

Only blue or violet or ultra-violet rays are capable of producing this phenomenon, and these may be obtained either by passing common white light from the sun, or the electric light, or the lime light, through a piece of blue or violet glass or through a solution of ammonia, sulphate of copper; or, better still, by producing a pure spectrum. The best effects are to be observed by using a prism of great dispersive power, like quartz.

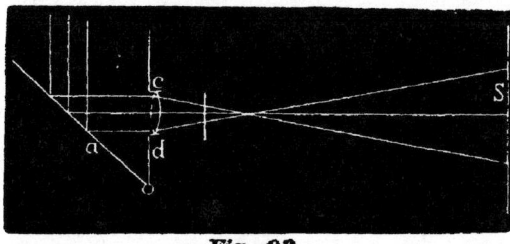

Fig. 93.

When colored glass is to be used to obtain the violet light, it suffices to place a lens of a foot focus near the orifice and the glass just in front of it. Fluorescent solids and solutions may then be examined at S. A piece of uranium glass or a solution of quinine in a test tube or bottle will exhibit this property so that many can see it at the same time. It will be well to use two bottles or beakers of clear glass, — one to contain pure water and the other the solution of quinine — and examine them side by side in this blue light. The fluorescence will then be more noticeable.

When artificial light is used in a lantern it will only be necessary to place the colored glass in front of the condenser, as if to project a picture upon it, and otherwise use the light as with sunlight.

Pictures are sometimes made of fluorescent material.

The outlines of flowers, butterflies, letters, etc., are drawn upon paper with a lead-pencil, and then painted with substances that exhibit different colors by fluorescence. When these pictures are used they may be pinned to the screen and the light allowed to fall upon them as before. Examine the pictures or other things by light transmitted through red, yellow, green, blue, and violet glass. The *kind* of light that induces fluorescent action will then be apparent.

When fluorescent substances are to be examined in the light of a solar spectrum they may be made to pass through it from the red towards the violet, and continuing beyond the visible part, for the ultra-violet rays are capable of powerfully exciting fluorescence in some substances. Stokes found this invisible spectrum that was competent to induce such action to be as much as three or four times the length of the visible spectrum.

The following substances manifest fluorescent action : —

Red Fluorescence,	Chlorophyl,
Orange "	"
Yellow "	Madder mixed with Alum,
Green "	Turmeric Stramonium and Night-shade,
	Brazil wood, Uranium glass, Thallene,
Blue "	Quinine — horse - chestnut bark, Petrolucene,
Purple "	Bichloranthracene.

These substances are generally prepared in solutions or decoctions for this purpose.

Chlorophyl may be prepared by boiling tea-leaves until water will remove nothing more, and then soaking them in hot alcohol. The chlorophyl will thus be extracted and the tincture is ready for examination.

A few pieces of the hulls of horse-chestnuts, or of the inner bark of the horse-chestnut tree digested half an hour in cold water, will be sufficient for this.

Alcoholic tinctures of madder, stramonium, night-shade, and Brazil wood will be needed.

A grain of sulphate of quinine may be put into a pint of pure water and shaken up occasionally. This substance is sparingly soluble in water. A little tartaric acid may be added to the water with advantage: the fluorescence will be more strongly marked.

A good method of exhibiting this is to have a rather large glass vessel containing pure water set in the path of the violet rays. Pour the quinine solution into it: opalescent clouds at once appear to form, though nothing is precipitated.

Thallene and anthracine are obtained from some of the products of the distillation of petroleum and coal tar, and are not in the market.

The Aurora tube and Geisler tubes when lighted by the electric spark may be used to obtain fluorescent effects. With the former, writing and drawings made with proper solutions, may be seen when such markings would be entirely invisible in common white light. Geisler tubes are often made to contain some pretty design in uranium glass, or there is some vessel containing a fluorescent solution surrounded with a jacket filled with some gas which gives a violet light like nitrogen.

A very beautiful effect is produced by exposing a number of highly fluorescent media to the flame of sulphur burning in oxygen in a dark room.

DOUBLE REFRACTION.

A piece of calc spar will be needed to show this. Its size is not very material, though the thicker it is the farther apart will the refracted figures be. It should have smooth faces, but the natural faces are often good enough to permit this phenomenon to be projected.

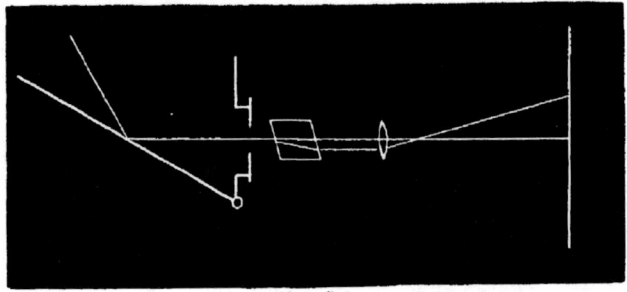

Fig. 94.

Make a hole a quarter of an inch in diameter through a bit of cardboard (unless you chance to have a diaphragm with holes of various sizes) and place it at the aperture; the small beam of light which comes through it should be directed horizontally upon the screen. Next place the piece of spar in front of it, and then project the hole with an object lens with a foot or more focus. The two spots will appear upon the screen, and if the spar be rotated the one spot will revolve about the other. Instead of the hole in a diaphragm, it will do as well to make a black spot upon a piece of glass and project it in the same way.

Either side of the spar may be used for showing this phenomenon.

A double-image prism may also be used with still better results as the images will be still further separated.

POLARIZATION OF LIGHT.

Plane-polarized light may be obtained in great quantity by using for a reflector in the *porte lumiere* a plate of glass blackened upon its back surface. Choose a piece of good window-glass, without bubbles or striæ, and paint it upon one side with lamp-black mixed in Japan varnish. It will be best to lay on two or three coats in order to completely cover the surface. Hold it between the eye and the sun, and all the uncovered and thin places can be seen; they should receive another coat of paint. This painted glass should be of the same size as the plane mirror in the *porte lumicre,* upon which it may now be placed and fastened by tying about them both a thread or stretching a ring of elastic cord over them. If the beam of light which is now reflected from the unpainted surface of this glass is sent through a double-convex lens and then received upon the screen, it will be seen to be

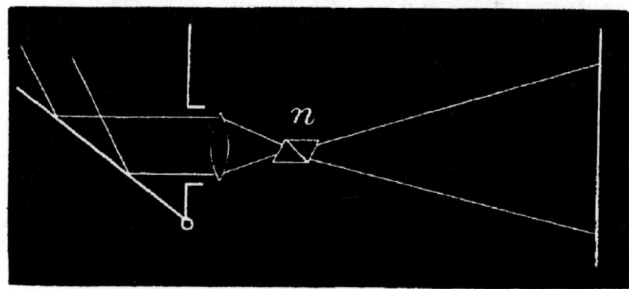

Fig. 95.

much less intense than the beam from the silvered mirror, but some of the most attractive experiments in the whole domain of physics are possible with this light.

A Nicol's prism *n* will be necessary, and the larger it is the better, but very good effects may be obtained with such small ones as come with the polarizing attachment to common microscopes. With one having a face three-quarters of an inch upon a side, everything essential can be shown to a large audience.

1. Place the prism at the focus of the lens so that all the light will pass through it. Now, if the prism be rotated upon the beam as an axis, the disk of light upon the screen will decrease in brightness until it is nearly or quite invisible; and if the prism be turned still further in the same direction the light will reappear and attain its maximum brightness when the prism has been turned ninety degrees from the position where the light disappeared.

2. Turn the prism so that the light is cut off from the screen; and then, holding it in that position, slowly introduce a thin sheet of clear mica between the lens and the prism. The light will reappear upon the screen from that transmitted by the mica. If the mica is as thin as the fiftieth of an inch, or less, the light may be colored a beautiful blue or green or red. Turn the mica round in its own plane, and these colors will appear in succession. Let the prism be rotated while the mica plate is held still, the same effects will be observed.

3. In the same manner experiments with thin plates of selenite may be tried.

4. Bring the lens forward so as to use it as an objective, and project a thin piece of selenite or of mica with varying thicknesses. Hold the prism in the focus as before. With each different thickness of the plates different colors will be transmitted which are often very beautiful indeed. If the pieces of these minerals are not more than an inch square, a larger lens may be used for a condenser, and then, with an objective of

four or five inches focus, project the piece in the same manner precisely as you would with the solar microscope. The prism must be held in the focus of the objective always.

5. Geometrical designs in mica.

Choose a thin plate of mica that is clear, and three inches or more square. Hold it in the polarized light and see if it presents lively colors; it will if it is thin enough. It ought not to exceed the fiftieth of an inch in thickness for best effect. When the tint appears uniform over

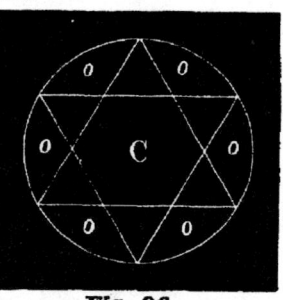

Fig. 96.

its whole surface, as it will if its thickness is uniform, it may have drawn upon it with a lead pencil such a figure as the accompanying one, and then trim to the edge of the circle with scissors. Afterwards, with a sharp pen-knife cut about one fourth of the way through the mica on all the lines; then with a needle point start to split the point on the edge. When a thin leaf has been raised a little between two points, carefully move the needle round the edge so as to separate the same thickness all around the circumference. Do not disturb the points of the star more than at the extreme point, just enough to keep the needle in the same layer. If the knife has cut through this layer that has been raised at the edge, the parts *o, o, o,* can be removed, leaving the six-pointed star a little raised above the surface *o, o, o.*

Examine this now with the polarized light, and the star will appear to be of one color and the cut-away parts of another. If the interior part *c* be removed it is very probable that that part will exhibit still another

9

color. If it does not, it is because the part removed had the same thickness as one of the others, or different from it by a wave length. Designs of any kind that fancy may dictate may be thus made upon sheets of mica. To project them plainly use an objective, as in 4, and place the Nicols prism in the focus of it.

Designs in selenite are still handsomer, and figures of birds, butterflies, flowers, and fruits may be bought in the market. Selenite is so brittle that a good deal of skill is needed to work it, and it would be tedious to a beginner. Such designs had better be purchased.

6. Unannealed pieces of glass when they have a regular form, as a square, a triangle, or a circle, make good objects to project by polarized light. They are generally a quarter of an inch or more in thickness, and an inch or two in diameter. Pictures of their appearance are often figured in works on physics. Pieces of thick glass, fragments of glass vessels, and glass stoppers of bottles often show double refractive power.

7. A good way to exhibit the development of this double refraction in glass is to take a piece of thick, plain glass and stand its edge upon a piece of iron, heated to redness, projecting the whole in the polarized beam with the prism in its place. As the glass is heated and strained, colors will develop upon the screen and arrange themselves symmetrically, depending wholly upon the external form of the glass.

8 It is convenient to have a piece of glass as much as a quarter of an inch in thickness and an inch square or more, that is annealed, and consequently gives no bands or colors. If this is strained by being pinched in a hand-vice, tufts of light or black brushes will be seen to start out from the place of pressure if the whole be projected.

9. A bar of glass half an inch thick, an inch broad, and six or eight inches long, may be gently bent with the fingers while held in place for projection. The strain induces double refraction, and that manifests itself by bands of light or dark, or color.

All of these should have their outline sharply projected by an objective of proper focal length.

10. A small crystal of Iceland spar, having its obtuse angles ground off and polished so as to present a surface as much as a quarter of an inch square, will present a beautiful series of rings and bars when projected.

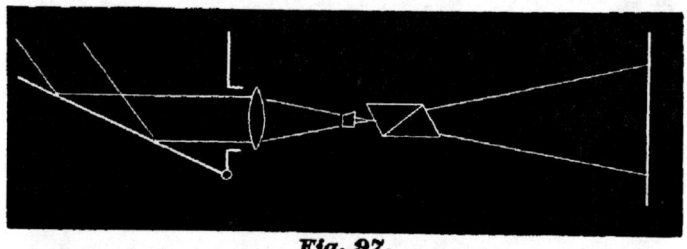

Fig. 97.

It will not be necessary to use an objective, but simply to put both crystal and prism in or so near to the focus of the condenser that as much light as possible may be transmitted through them. When in place let each in turn be rotated upon its axis, and observe the appearance and disappearance of the light and dark bands. At a distance of twenty feet from the prism the outer rings should be about four feet in diameter.

11. A crystal of rock candy, with parallel faces, and not more than the twentieth of an inch in thickness, will present another system of rings and bands. Project it in the same manner as the spar was projected.

12. A piece of a quartz crystal cut at right angles to its axis will, if projected in the same way as the last, exhibit colors upon the screen, which will vary as the

prism is turned. If it be put close to the prism there may appear a system of concentric circles about a uniform-colored field in the centre. The colors which this central field assumes when the analyzer is rotated are often superb.

Spectacle glasses that are usually called Brazilian pebble are made of quartz, and such will exhibit brilliant colors by projection in plane polarized light. This serves for a test of their genuineness, as glass will give no such effect.

Fig. 98.

Fig. 99.

13. The system of bands and colored curves seen in biaxial crystals is not easy to project, because the angles at which these are to be seen are so great. With some crystals of potassium nitrate it is possible to show both axes at a time with the same arrangement as was described for calc spar. A clear crystal about the quarter of an inch in diameter, and the twentieth of an inch thick, may answer for this. Such small crystals are usually mounted in a disk of cork.

Fig. 98 represents the double system of rings and brushes seen in a crystal of nitrate of potash, where the plane of its axes coincides with the plane of the polarizer; and Fig. 99 shows the appearance when the planes are slightly inclined to each other.

14. There are many minute crystallizations, such as are prepared for the microscope, that make fine objects when projected in polarized light. These objects may be prepared beforehand ; or the crystallization with the

accompanying development of polarization properties may be projected. It will be simply necessary to magnify the object by using a lens of short focus as in the former instruction for the solar microscope. The strip of clear glass, holding a drop of a saturated solution of the substance, the objective, and the Nicol's prism being put near the focus of a condenser of twelve to eighteen inches focus, that the specimen may be lighted as much as possible, and also have sufficient light transmitted.

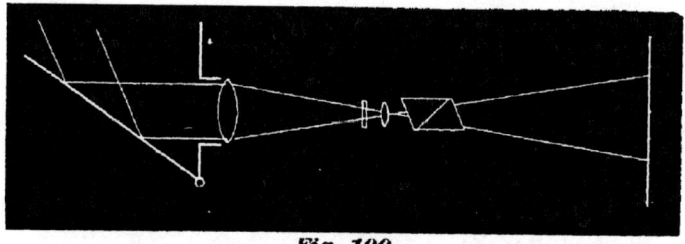

Fig. 100.

The following list of salts and other substances will be found to be beautiful objects for polarized light : —

Alum,
Borax,
Carbonate of Lime,
Carbonate of Soda,
Chloride of Barium,
    "   " Copper and Ammonia,
Chloride of Sodium,
Chlorate of Potash,
Citric Acid,
Nitrate of Bismuth,
    "   " Copper,
    "   " Potash,

Oxalate of Ammonia,
    "   " Lime,
Oxalic Acid,
Picric Acid,
Prussiate of Potash,
Salicine,
Sulphate of Copper,
    "   "   " and Magnesia,
Sulphate of Iron,
    "   " Soda,
    "   " Zinc,
Sugar,

Starch,

Tartaric Acid,

Urea,

Human hair,

Petals of flowers, as of the
Geranium,

Scales of Fishes.

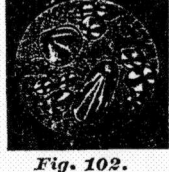

Fig. 101.   *Fig. 102.*

Fig. 101 represents the appearance of starch grains of the potato, as seen in common light with the microscope, and Fig. 102, the same seen by polarized light.

The following method of preparing double salts for examination with polarized light is given by Mr. Davies in the "Quarterly Journal of Microscopic Science":—

"To a nearly saturated solution of sulphate of copper and sulphate of magnesia add a drop on the glass slide, and dry quickly. To effect this, heat the slide so as to fuse the salts in its water of crystallization, and there remains an amorphous film on the hot glass. Put the slide aside and allow it to cool slowly. It will gradually absorb a certain amount of moisture from the air, and begin to throw out crystals. If now placed in the microscope, numerous points will be seen to start out here and there. The starting-points may be produced at pleasure by touching a film with a fine needle point so as to admit of a slight amount of moisture being absorbed by the mass of the salt."

A slide of salicine crystals makes a splendid object for such projection, and should be in every collection. Make a saturated solution of the crystals in distilled water, and place a drop carefully upon a slide that has been carefully cleaned. Evaporate over a lamp until it is dried to an amorphous mass. Upon cooling, a number of circular crystals will be formed with radiating forms between them. These circular crystals may be made larger and regularly disposed by touching the mass with a fine needle point when crystallization begins. Such ones will form about each point touched. Magnify such objects so much that these circular crystals will appear a foot in diameter upon the screen. The Nicol's prism will show each one with four arms that will turn about the centre of the crystal when the prism is rotated, while the radiating crystals will show as red, yellow, or purple brushes sweeping over the screen.

By inserting a sheet of transparent mica or thin selenite between the reflector and the object, a colored field will appear as a kind of background, upon which the minute crystals, such as chlorate of potash and oxalic acid, will appear more highly colored. The effect is usually to heighten the color.

THE DOUBLE IMAGE PRISM.

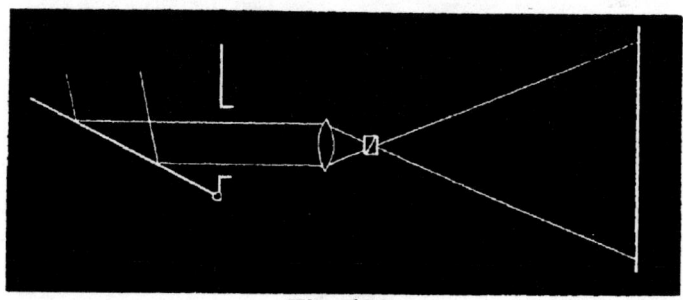

Fig. 103.

With a large lens project the image of the aperture upon the screen, the light being polarized by the blackened reflector. At the focus of the lens place the prism. Two images of the aperture will appear and overlap each other. Turn the prism on the beam as an axis; the images will turn about each other.

Place a thin piece of mica between the orifice and the lens. The two disks upon the screen will appear in complementary colors, save where they overlap, which will be white. Turn the prism as before; the colors of the two disks will change, always being complementary to each other.

Again, remove the reflector, and place the lens close to the orifice. Fix the prism near the focus so that a large part of the light passes through it; and then, with lens and Nicol's prism near it, project the images of objects placed close to the double-image prism. In this case the latter acts as a polarizer.

When large Nicol's prisms can be had, one of them may be substituted for the reflector upon the *porte lumiere.* The light passing through it will be polarized.

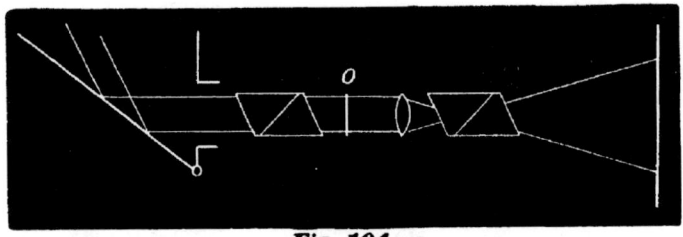

Fig. 104.

The object to be examined, *o,* may be placed near to it in front, then projected with any convenient lens, in the focus of which place the other Nicol's prism. This allows a long amount of light to be used, and is one method in use with lanterns. The only hinderance to the use of these larger prisms is the costliness of them.

All of these experiments may be performed with a lantern, with one of the more powerful lights. The usual method of polarizing the light is to have an elbow in front of the condensers that carries a series of plain glass plates inclined to the beam so that it meets it at the polarizing angle of glass. Part of the light is transmitted and is absorbed by a piece of black cloth. The light that is reflected is sufficiently well-polarized for all purposes of demonstration; and such a beam may be treated in every way like the beam from the *porte lumiere* and with like results.

DIFFRACTION.

Reflect the beam from the *porte lumiere* through a slit like one for showing the Fraunhofer lines. It ought not to be more than one sixteenth of an inch wide. Receive this beam, without magnifying it, upon a second slit in a screen at a distance of four or five feet from the first slit. Make the room as dark as possible, and then hold a sheet of white paper behind the second slit anywhere from a few inches to several feet. Colored fringes will appear on each side of the central line, with a series of alternate black and white bands or lines. These may be received upon a screen twenty feet away, when they should have a united breadth of a foot or more, but the light is necessarily very weak. A lens does not improve them very much.

With a piece of perforated paper or tin or lace, or still better, with an eidotrope, which consists of two disks of perforated tin made to revolve in opposite directions, like the chromatrope, a very beautiful exhibition of the phenomenon of diffraction may be given in the following way : —

Take two large, short, focus lenses, such as form the condensers in Marcy's sciopticon. Place one close to the opening to the *porte lumiere,* as shown in the figure. The second one may be put so far in front of the other

lens that the beam is again crossed in front of it, and
the disk upon the screen is of the desired size, six or
eight feet in diameter. Now introduce the perforated
paper or the eidotrope at the place marked e. The

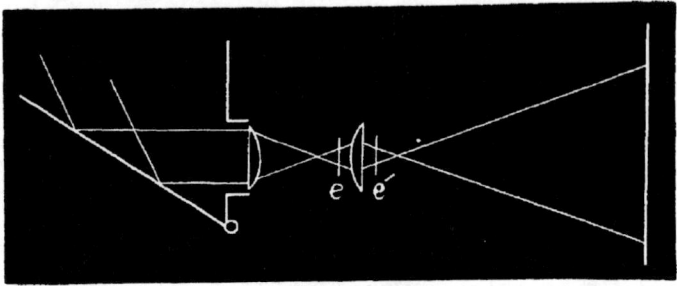

Fig. 105.

screen will appear covered with minute spectra, as
each hole will form one or more spectra ; but if the
paper be held at é, between the lens and the screen, the
projection becomes very gaudy and symmetrical. If it
be the eidotrope, turn it while held in that place, and
the colors will change and will rival the colors produced by polarized light. Try the effect of a comb,
of wire gauze, of the fingers, or other objects. Very
curious and interesting appearances will appear upon
the screen.

If one has a piece of glass finely ruled with a dia-
mond, it may be projected as is any object with the
porte lumicre, and the diffraction spectra will appear
upon the screen. With plenty of light for the projection, and with the room otherwise well darkened, a
number of the Fraunhofer lines may be seen in these
spectra.

Again, let a concentrated solution of alum or camphor be poured upon a glass plate, and allowed to dry
rapidly, so as to cover it with a crust. Put it in the
focus of a lens with a short focus, and a series of

halos will be seen by placing a small screen but a foot or two from the glass. Fine rulings upon blackened glass will in the same place give fine colors. These rulings may be as coarse as fifty to the inch ; the finer rulings will answer better. The rapidly-diverging rays necessitate the placing of the screen close to the plate, else the colors will be too faint.

PERSISTENCE OF VISION. — THE STROBOSCOPE.

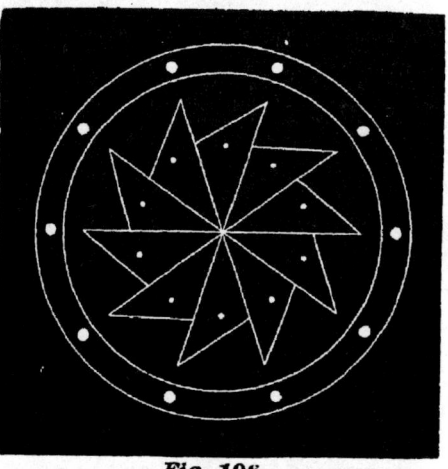

Fig. 106.

Let a disk a foot in diameter be cut out of any convenient material,—tin, copper, zinc, or pasteboard. Near the periphery cut out a number of holes at equal distances apart, — ten or twelve will be enough. They may be cut half an inch is diameter. This disk is to be put upon a rotator like the one used to show the Newton's disk, and may now be placed so that the focus of the condenser with the *porte lumiere* will be in the holes as the disk revolves, as in Fig. 106. This permits the light to pass to the screen only when the holes are at the focus, at which time a powerful beam will pass and is immediately cut off. With such a fixture a very great number of amusing and instructive experiments may be made.

1. While one person turns the stroboscopic disk let another one stand in front of the screen and swing his arms, or move his body rapidly sideways, or make

low courtesies. To spectators he will appear to have a dozen arms or bodies. There will also be as many shadows upon the screen.

2. Make a wheel to turn in front of the screen, the

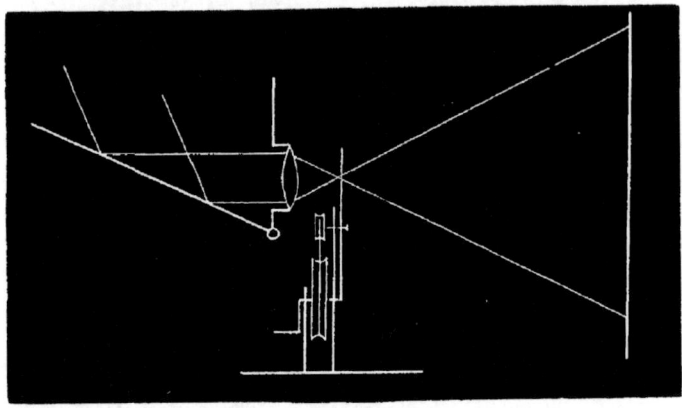

Fig. 107.

larger the wheel the better. A buggy wheel or old fashioned spinning-wheel make good objects. Let both disk and wheel be turned at the same time. The appearance of the wheel will depend upon its velocity. It may be made to appear as if standing still or moving slowly forward or backward, or as if it had a multitude of spokes.

3. While the wheel is turning a little in front of the screen, look *through* the wheel at the shadow of it. Some remarkable curved lines will appear to group themselves about the axles of the wheel and its shadow.

4. If two wheels of the same size are made to turn one in front of the other while they are in this intermittent light, curious curves and fixed straight, light or dark lines and mixed, changing paths can be seen, according to the position the spectator has with reference to the wheels.

5. If a small wheel but two or three or four inches in diameter, and toothed like a cog-wheel in a clock, be placed within a foot or two of the disk, and so that its shadow will fall upon the screen, its shadow will not only be much magnified, but the motions of the wheel will appear as with the larger one, number 1.

6. Let large disks three or four feet in diameter be made, having various symmetrical figures painted upon them. When the disks are revolved, many curious motions may be simulated : as of a girl jumping rope, a man sawing or chopping wood, boys playing leap-frog, a man opening and shutting his eyes and mouth, wind-mill with sails turning, etc. Such designs generally come with the toy called the thaumatrope, made to look through into a mirror while turning. These may be copied upon large sheets of pasteboard and rotated in any convenient way. The turning-table may be made to answer for this.

7. Another good and very pretty application of this is to have a large star with five or six points made, and the alternate points colored with different tints, as red and blue. When such a disk is revolved in this intermittent light it may appear to stand very still, or to slowly revolve forward or backward, while its points may be doubled or tripled or quadrupled, and its colors will apparently overlap and give the tints proper to the mixture.

8. Such pictures as are sold with the thaumatrope may be fastened to the front of the disk containing the holes through which the light passes, as is represented in Fig. 107, and after the light has passed through the disk, it may be reflected upon its face by a small mirror, *m* (Fig. 108), and can thus be seen very well if the light be strong. When used in this way the disk

may be made very much larger, as much as two or three feet in diameter, and the number of holes in-

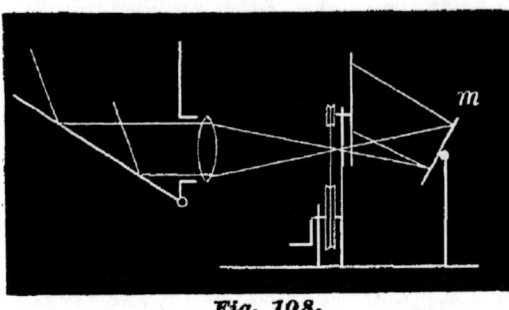

Fig. 108.

creased. By removing the mirror *m* a little farther away the beam can be reflect-ed so as to cov-er the whole face of the disk.

A small toy steam-engine, such as may be bought for a dollar or two, may have a light paper disk fitted for it to turn, but if sunlight be used, care must be taken lest it take fire in the focus of the sun's rays.

An oxyhydrogen lantern may be used for such work. The objective will need to be removed, and the perfo-rated disk placed so that the most of the light goes through the holes when they are in position, and the unused light cut off from entering the room by black cloth or some other provision. Otherwise it will be used just as with sunlight.

THE CHROMATROPE.

This instrument consists of two disks of glass so mounted that they may be rotated in opposite direc-tions. Various designs are painted upon the disks, and fine effects may be obtained by projecting them in the ordinary way with the lantern or the *porte lumiere.* If instead of using disks of glass, disks are made of wire gauze, perforated tin, or paper or lace, very curious interference figures are produced, and this form is called the eidotrope.

The accompanying figure represents a chromatrope with an arrangement for quickly replacing one disk by another of different pattern. Rotation is given by friction pulleys. With this form there is a disk with the so-called seven primary colors to illustrate Newton's theory of colors,

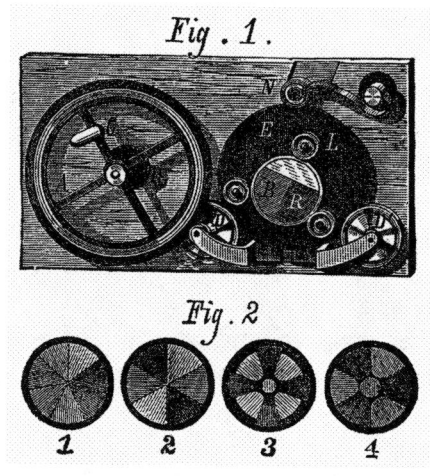

Fig. 1.

Fig. 2

1   2   3   4

one to illustrate Brewster's theory, two to illustrate Young's theory, and a chameleon top, designed by President Morton, of Stevens Institute, Hoboken.

The effects with all the forms of chromatropes are due to persistence of vision.

Interesting subjective effects may be observed by projecting in the ordinary way bits of colored glass an inch or two square, so as to have upon the screen a large patch of color with a boundary of white light. The eyes must be fixed attentively upon the colored patch for about half a minute, when the colored piece must be quickly removed, the eyes to be kept meanwhile upon the screen. To prevent the eyes from unconsciously wandering while looking, it will be found advisable to pin a large black button or a piece of black paper to the screen in the middle of the disk. This is to be kept in the centre of vision. The effects observed will of course depend upon the color upon the screen, and the sensitiveness of the eyes for various colors. Generally, after looking steadily at a given color, and the disk is made suddenly white, the outline of the colored part will be seen in a color complementary to the one looked at first. Thus, if a square red glass should be projected the residual image would be a square green one. If a blue one was projected its complementary image would be orange, and so on. A great variety of such effects are obtainable with various colored pieces of glass, or of films of gelatine, by projecting them singly, in juxtaposition, or superposed.

Let disks of white cardboard a foot or two in diameter have partial sectors painted black, with india ink, so that the white and black parts alternate four or five times in the circumference. This is to be rotated while a powerful beam of light falls upon it. The persistence of some of the elements of white light being greater than of others, the disk will appear of various colors ; purple, green, and yellow being generally well developed.

HEAT. — AIR THERMOMETER.

A bulb blown upon one end of a small glass tube, five or six inches long, answers for this experiment. A drop of colored water can be made to enter the tube by first heating the bulb a little by holding it in the fingers with the open end of the tube a little below the surface of the water. A bubble or two of air will be expelled, and the fingers may be removed from the bulb. As it cools a drop will be driven into the tube, and with a little painstaking it can be brought to any required place by cooling or heating the bulb. These movements can be shown with the *porte lumiere* and a single lens, as shown in Fig. 17, or it can be put in front of the condenser of the lantern. A touch of the finger will heat the bulb sufficient to cause the drop to rise in the tube, and it may be made to descend by simply blowing upon the bulb, or by dropping a little water or ether upon it.

Many of the pieces of apparatus for illustrating the expansion of metals by heat are so small that they may be readily projected. Thus Gravesand's Ring, Pyrometers, etc. The latter may have a small bit of mirror fastened to the end of the index, and the light so arranged that as the index rises, the beam will move upward. A rise in temperature of only a few degrees can be then shown, and the alcohol flame may be dispensed with; the warmth of the hand or a little hot water answering the purpose.

FORMATION OF CLOUDS.

The condensation of liquid in the form of vapor into minute globules and in the production of a shower of rain may be very well illustrated and projected for class purposes in the following manner:—

Place about an ounce of Canada balsam in a Florence flask and make it boil. At the top of the flask clouds of globules of turpentine will be seen hovering about, altering in shape very much like sky clouds, and the globules are large enough to be visible by the naked eye. If a cold glass rod be gradually introduced into the flask these clouds may be made to descend in showers. *Lawson Tait in Nature.*

Another: Take a flask of one or two litres capacity; rinse it out with distilled water, and attach to the neck a cork and glass tube of about twenty or thirty centimetres length. Place the glass tube in the mouth and exhaust, when a dense cloud will be formed; then on blowing into the flask the cloud disappears. The cloud may be produced and dissolved as often as wished, and if a beam from the oxyhydrogen light be sent through the flask the experiment becomes very effective.   *C. J. Woodward in Nature.*

MAXIMUM DENSITY OF WATER.

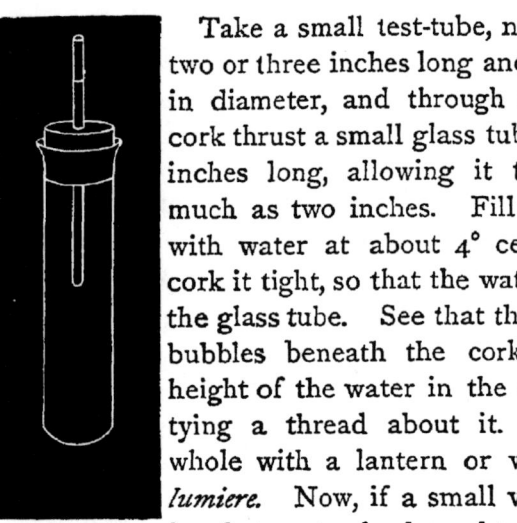

Take a small test-tube, not more than two or three inches long and half an inch in diameter, and through a tight-fitting cork thrust a small glass tube about three inches long, allowing it to project as much as two inches. Fill the test-tube with water at about 4° centigrade and cork it tight, so that the water will rise in the glass tube. See that there are no air bubbles beneath the cork. Mark the height of the water in the small tube by tying a thread about it. Project the whole with a lantern or with the *porte lumiere.* Now, if a small vessel contain-

Fig. 109.   ing hot water be brought up under the test-tube so that the latter dips in it, the expansion of the water will be indicated by the rise of the water in the tube, and the latter will overflow if it be sufficiently heated. Now, bring up under it in the same way a freezing mixture of ice and salt, or a mixture of equal parts of cold water and nitrate of ammonium. The water will contract in volume to its minimum, which should be indicated by the thread; then it will again expand until it freezes, the expansion again causing the water in the tube to overflow. The freezing mixture should be stirred constantly to hasten the work.

THE GALVANOMETER.

As many of the experiments in heat require the thermo-pile and galvanometer, the latter is treated of in this place rather than with electrical experiments.

In the "American Journal of Science," Vols. II, III, V, IX and X, are given several ingenious arrangements for projecting the movements of a galvanometer needle, and if one desires to do extremely accurate work before an audience he will do well to obtain some one of these forms. If, however, it is desirable only to exhibit qualitatively and with no great degree of precision the relation of heat to electricity, or the law of the galvanometer, etc., the following method will be found to answer, with the advantage of being extemporized in a few minutes: Make a *flat coil* about an inch square, of rather fine-covered copper wire having the ends of the wire a foot or more in length. Upon one side of this coil stick a bit of beeswax as large as a small marble, and through both wax and coil thrust half of a fine cambric needle. Press the wax down upon the middle of a piece of glass four or five inches square, and then holding the plate horizontal, suspend upon the needle point a small compass needle an inch or two long. This is now ready to place upon the upper condenser *c* (Fig. 27) of the vertical attachment and then be projected. If a current from a battery or a thermo-pile be sent through the coil, the needle will be deflected. The needle will of course point towards the north, and that place will easily be noted upon the screen as zero. A small magnet brought into the neighborhood will serve to bring the north pole of the needle to any required place. If a circle with inscribed degrees should be drawn upon the glass by either of the methods described upon pages 31 or 32, the movement of the needle can be noted in degrees. If the needle is too short to reach the numbers upon the glass, it can have a fine straight bristle made fast to its ends with a little mucilage.

With the thermo-pile connected with the galvanometer, the sensitiveness of the former may be shown by presenting the hand to one face of it, or it may be breathed upon or blown upon with a common hand bellows. Let fall a drop of water, of ether, and of alcohol upon the face. The evaporation cools it.

The heat generated by percussion may be exhibited by providing a leaden bullet which should have at first the same temperature as the thermo-pile, which may be known by putting it upon the pile, handling it with a pair of small tongs. It should not move the needle. Then strike it once with a hammer so as to indent it considerably, and with the tongs quickly put it again upon the face of the pile. It will indicate a higher temperature.

The heat generated by friction may be shown by rubbing a stick upon the floor and then bringing it to the pile as in the other case.

See Tyndall's work on Heat for a method of showing heat from the crystallization of sodium sulphate. The same thing may be shown with the air thermometer sunk into the solution, which may be projected with lantern or *porte lumiere* by preparing the solution in a beaker, fixing the air thermometer in it with a drop of colored water in it, and projecting the whole upon the screen by means of a large lens. The crystallization itself will be seen, as well as the manifested heat, when it reaches the bulb of the thermometer.

Mix in a test-tube resting upon the face of the thermo-pile, a few drops of water and sulphuric acid about equal parts: the heat evolved will illustrate the origin of heat from chemical reaction.

A few crystals of nitrate of ammonium in a test-tube may have an equal bulk of water poured upon them; the cold produced is from the absorption of heat during liquefaction.

Interpose between the source of heat and the thermo-pile various things, such as rock-salt, a solution of iodine in bisulphide of carbon, glass, crystals of various kinds, tubes filled with gases and vapors of various sorts. Also, project a solar spectrum with a part of the same beam that projects the galvanometer by the method described upon page 112. Move the thermo-pile through the various colors, and note the degree indicated by the galvanometer, particularly beyond the red end of the spectrum. The thermo-pile should be placed where the Fraunhofer lines are seen best upon a small screen placed temporarily to receive it.

Many experiments on this subject will be found in Tyndall's work on Heat, which one will find himself able to repeat with satisfaction.

CALORESCENCE.

Let the light from the *porte lumiere*, or from the electric or lime light, be sent through a vessel containing bisulphide of carbon in which some iodine has been dissolved: the solution will be jet black and will stop every light ray, but will permit the rays of greater wave length to freely traverse it. A lens may now be interposed and the obscure rays treated in every way like luminous rays. With a very powerful beam platinum foil may be raised to incandescence in the focus of the lens, and with a less powerful one pieces of wood and paper may be ignited.

A transparent solution of common alum is opaque to the same rays that are so easily transmitted by the iodine solution.

A test-tube filled with water placed at the focus of the obscure rays in a minute or two may be made to boil; an air thermometer will scarcely be affected at that place.

MAGNETISM.

With the vertical attachment to the lantern the phenomena of magnetism are best shown.

1. Have two or three small magnetic needles mounted upon needle points thrust through pieces of cork, so as to turn freely. Place one upon the upper face of the condenser to the vertical attachment, and project it sharply upon the screen. A piece of iron or another magnet brought into its neighborhood will disturb it, and every motion will be plainly noticeable as well as the direction of the exciting body.

2. Place *two* of these needles near to each other, but not so near as to touch, and give to one of them a twirl so that it revolves upon its support. It will soon set the other revolving and it may be stopped itself after setting the second one going, and afterward be again started while the other one stops.

3. Place a third, quite small one not more than half an inch long in the neighborhood of the other two, and again set the one whirling.

4. The magnetic phantom.

Lay a small magnet an inch or two long upon the upper condenser; and upon the magnet lay a piece of clear glass three or four inches square. Project the magnet, and then scatter from a small sieve, or gently with the thumb and finger, fine iron filings upon the glass. The filings will arrange themselves in the familiar lines called the magnetic phantom, and the whole being magnified to ten feet or more in diameter makes a very striking picture.

5. The elongation of an iron rod when strongly magnetized, may be shown by placing a small helix around the iron rod of the common pyrometer made for showing the longitudinal expansion of a rod by heat. To the end of the index finger that sweeps over the quadrant affix a small bit of plane mirror not more than one fourth of an inch square. So adjust the light to this small mirror that the reflection from the latter will fall upon the most distant part of the room ; the farther away the better. When the current of electricity is sent through the helix the rod will be slightly elongated, but the slight tilting of the mirror may become a displacement of two or three inches at a distance of thirty feet.

DIAMAGNETISM.

The electro-magnet for demonstrating diamagnetic phenomena need not be over three or four inches in length, and the poles an inch apart. Objects to be tested may be suspended by a thread between the poles, and the whole projected either in a beam of parallel rays or in front of the focus of a lens. In the latter case the whole will be seen in profile, but perfectly distinct. The following experiments may be projected with such a magnet if a battery of three or four cells be used : —

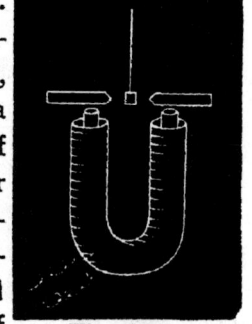

Fig. 110.

1. Suspend oblong pieces of various metals half an inch in length, and note whether they set themselves equatorially or axially between the poles. Iron, nickel, platinum, bismuth, antimony, zinc, tin, lead, silver, copper, alum, glass, sulphur, sugar, bread, paper, charcoal, are good substances to experiment with.

2. Suspend a cube of copper between the poles, and twist the thread so that the copper will rotate rapidly by torsion. It will quickly be brought to rest when the current is made to pass.

3. Fill small very thin tubes with liquids, and suspend them in the same manner. Try solutions of iron, cobalt, water, alcohol, turpentine, and salt.

4. Place the magnet upon the upper condenser of the vertical attachment, and upon its poles place a watch-glass containing a little water or sulphuric acid; project the water in the watch-glass, and notice the distribution of light upon the image of the water. Now complete the circuit. The water will change its form slightly and the light will be differently refracted, thus making it quite visible. Salts of iron or nickel will scatter the light like a concave lens.

5. Hold the flame of a candle between the poles.

6. Blow small soap bubbles with oxygen and with illuminating gas, and hold them as close to the poles as possible or drop them so they will rest upon both.

7. Heat a coin and place it just beneath the poles, and then drop a piece of iodine upon the coin. The heat will volatilize the iodine, and the purple vapor will be repulsed.

ELECTRICITY.

Most of the experiments in electricity which can be shown by projection require the use of the galvanometer, such for instance as give evidence of the existence of electrical currents, their direction and strength. These will only need the arrangement already described under the head Galvanometer. For other experiments, such as that of the electric light, there will be needed some one of the many fixtures for holding the carbons to be ignited. If this can be put into a lantern the carbons may be projected at once upon the screen by removing the objective and drawing the carbons back until the image appears plainly upon the screen. This image will be made much sharper by putting a diaphragm with about an inch aperture over the condensers, which in this case serves for an objective.

For the projection of spectra precisely the same conditions need to be observed as for the lime light: — Some regulator in the lantern, a slit in the focus of the condensers, an objective to project the slit and the prism in the focus in front of the objective. The spectrum of metals is easy with this arrangement. Make a small cavity in the end of the lower carbon stick, and place a small bit of the metal whose spectrum is wanted in it; then bring down the upper carbon upon it so as to complete the circuit and then raise it a little, the metal will be at once fused and volatilized, emitting its characteristic light, which will appear upon the screen as bright bands. Silver, copper, zinc, iron, and mercury give good spectra among the more common elements.

For the successful working of this method of spectrum analysis, not less than forty cells will be needed, and fifty are decidedly better than forty.

DECOMPOSITION OF WATER.

This is effected by sending a current of electricity from three or four cells through water that has been slightly acidulated by the addition of a little sulphuric acid. The terminals of the wires in the water are usually made of strips of platinum to prevent other chemical reactions from taking place. For projection, an excellent way is to insert two test-tubes filled with the acidulated water, and introduce them into the tank already described, having previously fixed the two platinum terminals through the rubber bottom as

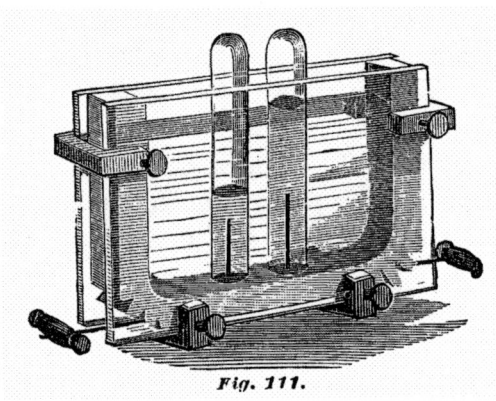

Fig. 111.

shown in Fig. 111. When the current is sent through these wires the bubbles will rise rapidly and soon fill the hydrogen tube. This tank is of course to be projected in the ordinary way, either with lantern or *porte lumière*, in which case the bubbles will appear very large and the water will appear to be in great commotion.

In place of water fill the tank with a solution of acetate of lead, and without the test-tubes project the tank and make connection with the battery of two or three cells as before: the crystallization of the lead will at once begin and rapidly grow upon one of the terminals; reverse the current, and the formed crystals will dissolve while others will grow upon the other terminal. The same thing can be done still better by filling the horizontal tank for the vertical attachment with the solution of lead acetate, and then bending a piece of platinum wire or of tin wire around the interior of the tank. Then, on inserting another wire at the centre of the solution, and making connection with two or three cells so as to make the centre wire the negative and the hoop the positive pole, a beautiful growth of metallic crystals will shoot out from the centre and spread out over the entire field. In place of the solution of lead use a strong solution of the bichloride of tin, using a tin hoop in the solution. Crystals of tin will shoot out and appear in great beauty.

These solutions in the horizontal tank should not be more than an eighth of an inch deep.

HEATING BY THE CURRENT.

Make a small coil of platinum wire, and thrust the ends of the wire through the rubber of the tank, as

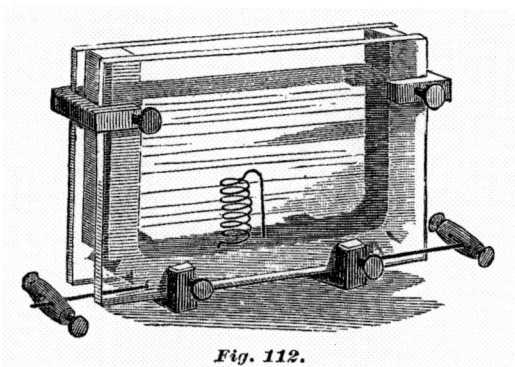

Fig. 112.

shown in the engraving, Fig. 112. Fill the tank with water, and having projected the whole, send the current through the wire. If the current is sufficiently great the wire coil will be heated at once, and a convection current will at once show itself in the water, the heated water next to the wire rising rapidly to the top. The effect will be still more marked if a drop or two of some one of the aniline dyes be let fall from the surface over the wire. Its greater density will carry it at once to the bottom; but when the current is sent through the wire, the movements in the water will be rendered very plain. The bichloride of tin or the sulphate of zinc will also answer the same purpose.

CHEMISTRY.

Most of the chemical reactions that are usually exhibited before classes in the recitation or lecture-room can be shown in a much more satisfactory way by means of the apparatus for projection than in the ordinary way. The method is moreover both cheaper and easier; cheaper, because each experiment requires but a few drops of the substance in a test-tube or the tank, instead of the large quantity necessary for many to see at once, and easier, because the preparation needed for experiments upon an extended scale is always tedious and tiresome. One who uses the tank (Fig. 20) for the first time for projection, say of silver, in a solution as dilute as two or three drops of the nitrate to the tank full of water, will be surprised at the prodigious precipitation brought about by the addition of a single drop of hydrochloric acid introduced upon the end of a glass rod. Great heavy clouds roll and tumble about upon the screen, looking as though they might weigh tons.

ACIDS AND ALKALIES.

Nearly fill the tank with water and add a few drops of blue litmus solution; then dip a glass rod into a weak acid solution of any convenient kind and gently stir the litmus solution with it: it will turn red in the neighborhood of the rod. After washing the rod, dip it into an alkaline solution of ammonia or potash, and again stir the solution in the tank. Blue clouds will form in the red sky upon the screen until the whole is again a beautiful blue.

In place of litmus solution use a solution made by boiling purple cabbage. Acid turns this red, and an alkali turns it green. Such changes may be effected a number of times in succession in the same solution.

Nearly fill the tank with sulphate of soda, in which is put either litmus or cabbage solution to color it; the latter is the best. After projecting it as a blue solution dip the terminals of a battery of three or four cells into it. Decomposition will begin and the acid and alkaline reactions will be observed about the poles.

REACTIONS AND PRECIPITATION.

Fill the clean tank nearly full of pure water and add a drop or two of a solution of nitrate of silver and stir it well. Then dip the glass rod into very dilute hydrochloric acid. Very dense clouds of chloride of silver will form and fall to the bottom of the tank. Add a few drops of strong ammonia water, and the cloudy solution will again become clear.

A small piece of carbonate of lime or of soda placed in the tank containing a very dilute solution of hydrochloric acid gives up its carbonic acid in apparently large quantities.

To water made slightly acid, add enough litmus solution to turn it red and project it ; then drop a lump of carbonate of ammonia into it. It will dissolve rapidly with effervescence, and the solution will be made blue about the crystal, and if there is enough of it the whole solution will ultimately become blue.

The gradual solution of substances in water may be nicely shown by filling the tank with pure water and dropping a crystal of alum or sulphate of zinc or sulphate of copper into it. Where the substance is dissolved the solution will be denser, and its refractive powers changed, which will be manifest by gently stirring it with a glass rod.

A dilute solution of copper sulphate may be placed in the tank. With a pipette, force into the solution some ammonia water : A dense precipitate will at first be formed, which will afterwards be dissolved if ammonia enough has been added, leaving the solution a beautiful blue color. A few drops of sulphuric acid will reproduce the precipitate and destroy the color ; and when the solution again becomes clear, a few drops of ferrocyanide of potassium added will produce a brownish-red bulky precipitate, which will present a fine appearance upon the screen.

In like manner all of the characteristic reactions of inorganic chemistry may be projected, and often with much less expenditure of materials than would be used if large vessels were employed to demonstrate the same things. One who has projected a number of these phenomena will find no difficulty in projecting any reaction that may be observed in a test-tube.

Pictures of chemical apparatus, of processes, etc., will be very convenient for projection when instruction is given in chemistry.

UNIFORM WITH THE "BOOK OF AMERICAN EXPLORERS."

YOUNG FOLKS'
ℌistory of the ⅄nited ⅀tates.

BY
THOMAS WENTWORTH HIGGINSON.

Square 16mo. 380 pp. With over 100 Illustrations. Price $1.50.

The theory of the book can be briefly stated: it is, that American history is in itself one of the most attractive of all subjects, and can be made interesting to old and young by being presented in a simple, clear, and graphic way. In this book only such names and dates are introduced as are necessary to secure a clear and definite thread of connected incident in the mind of the reader; and the space thus saved is devoted to illustrative traits and incidents, and the details of daily living. By this means it is believed that much more can be conveyed, even of the philosophy of history, than where this is overlaid and hidden by a mass of mere statistics.

"Compact, clear, and accurate. . . . This unpretending little book is the best general history of the United States we have seen." — *The Nation.*

"The book is so written, that every child old enough to read history at all will understand and like it, and persons of the fullest information and purest taste will admire it." — *Boston Daily Advertiser.*

"It is marvellous to note how happily Mr. Higginson, in securing an amazing compactness by his condensation, has avoided alike superficiality and dulness." — *Boston Transcript.*

AS A TEXT-BOOK IN SCHOOLS.

One of the most successful teachers in Boston says, "I am confident that the text-book has proved itself as reliable and comprehensive as it certainly is suggestive and entertaining. I know no book more helpful in promoting that crystallizing process in the student's own mind by which the accessories and details group themselves around the main facts and ideas of the narration. On this account, it is equally valuable to teachers and scholars, to the examined and the examiners."

This work has been translated into German, and has been received with marked favor. The Leipsic literary correspondent of the "New-York Staats-Zeitung" says, that, in its German version, it is pronounced exceedingly interesting (*höchst anziehende*); and predicts that it will inspire universal delight (*allgemeine Beliebtheit*) in German readers.

The Berlin "International Gazette" says, "Mr. Higginson has executed his task in a very clear and lucid manner, not making use of any hard aphorisms, so puzzling to the young, but placing himself on their level, and explaining every thing in so easy and gentle a manner, that he must be a very dull or a very perverse scholar, who does not find his attention riveted."

** Sold by all Booksellers, and sent by mail on receipt of price.

LEE & SHEPARD, Publishers,
41 FRANKLIN STREET, BOSTON.

JUST READY.

A New Work by the Author of the Young Folks' History of the United States.

YOUNG FOLKS'
BOOK OF AMERICAN EXPLORERS

BY

THOMAS WENTWORTH HIGGINSON.

Uniform with the Young Folks' History of the U. S. One vol. Fully illustrated.
Price, $1.50.

The YOUNG FOLKS' BOOK OF AMERICAN EXPLORERS is as distinctly a "new departure" in our historical literature as was its predecessor, the "Young Folks' History of the United States." The "Book of American Explorers" is a series of narratives of discovery and adventure, told in the precise words of the discoverers themselves. It is a series of racy and interesting extracts from original narratives, or early translations of such narratives. These selections are made with care, so as to give a glimpse at the various nationalities engaged, — Norse, Spanish, French, Dutch, English, etc., — and are put together in order of time, with the needful notes and explanations. The ground covered may be seen by the following list of subjects treated in successive chapters : — The Traditions of the Norsemen ; Columbus and his Companions; Cabot and Verrazzano ; The Strange Voyage of Cabeza de Vaca ; The French in Canada ; Hernando de Soto ; The French in Florida ; Sir Humphrey Gilbert ; The Lost Colonies of Virginia ; Unsuccessful New England Settlements ; Captain John Smith in Virginia ; Champlain on the War-Path ; Henry Hudson and the New Netherlands ; The Pilgrims at Plymouth ; The Massachusetts Bay Colony.

Besides the legends of the Norsemen, the book makes an almost continuous tale of adventure from 1492 to 1630, all told in the words of the explorers themselves. This is, it is believed, a far more attractive way of telling than to rewrite them in the words of another ; and it is hoped that it may induce young people to explore for themselves the rich mine of historical adventure thus laid open.

LEE & SHEPARD, *Publishers, Boston.*

Made in the USA